simple methods for aquaculture

TOPOGRAPHY
for freshwater fish culture
topographical tools

Text: A.G. Coche
Graphic design: T. Laughlin

FOOD AND AGRICULTURE ORGANIZATION OF THE UNITED NATIONS
Rome 1988

David Lubin Memorial Library Cataloguing-in-Publication Data

FAO, Rome (Italy)
    Topography for freshwater fish culture: topographical tools.
    (FAO Training Series, no. 16/1)
    ISBN 92-5-102590-8

    1. Topography   2. Fish culture
    I. Title   II. Series

    FAO Code: 44   AGRIS: P31   M12

P-41

ISBN 92-5-102590-8

THE AQUACULTURE TRAINING MANUALS

The training manuals on simple methods for aquaculture, published in the FAO Training Series are prepared by the Inland Water Resources and Aquaculture Service of the Fishery Resources and Environment Division, Fisheries Department. They are written in simple language and present methods and equipment useful not only for those responsible for field projects and aquaculture extension in developing countries but also for use in aquaculture training centres.

They concentrate on most aspects of semi-intensive fish culture in fresh waters, from selection of the site and building of the fish — farm to the raising and final harvesting of the fish.

The following manuals on simple methods of aquaculture have been published in the FAO Training Series:

Volume 4 — Water for freshwater fish culture
Volume 6 — Soil and freshwater fish culture
Volume 16/1 — Topography for freshwater fish culture
Topographical tools
Volume 16/2 — Topography for freshwater fish culture
Topographical surveys

The following manuals are being prepared:
Pond construction for freshwater fish culture
Management for freshwater fish culture

FAO would like to have readers' reactions to these manuals. Comments, criticism and opinions, as well as contributions, will help to improve future editions. Please send them to the Senior Fishery Resources Officer (Aquaculture), FAO/FIRI, Via delle Terme di Caracalla, 00100 Rome, Italy.

HOW TO USE THIS MANUAL

The material in the two volumes of this manual is presented in sequence, beginning with basic definitions. The reader is then led step by step from the easiest instructions and most basic materials to the more difficult and finally the complex.

The most basic information is presented on white pages, while the more difficult material, which may not be of interest to all readers, is presented on coloured pages.

Some of the more technical words are marked with an asterisk (*) and are defined in the Glossary on page 325.

For more advanced readers who wish to know more about topography, a list of specialized books for further reading is suggested on page 330.

CONTENTS

CONTENTS, continued

LIST OF TABLES

LIST OF FIGURES

1 GENERAL BACKGROUND

10   What is topography?

Topography is the science of measuring the earth and its features, and of making maps, charts and plans to show them. These features may be **natural**, such as plains, hills, mountains, lakes, streams, rocks or forests. They may also be **man-made**, such as paths, roads, buildings, villages or fish ponds. A topographical map can also show the **slope** of the ground. It can show not only which places are high and which are low, but also how steep the land is between high and low points.

Sometimes topography is also called **surveying**. A person whose profession is making topographical measurements and recording them on maps, charts and plans is called a **surveyor**.

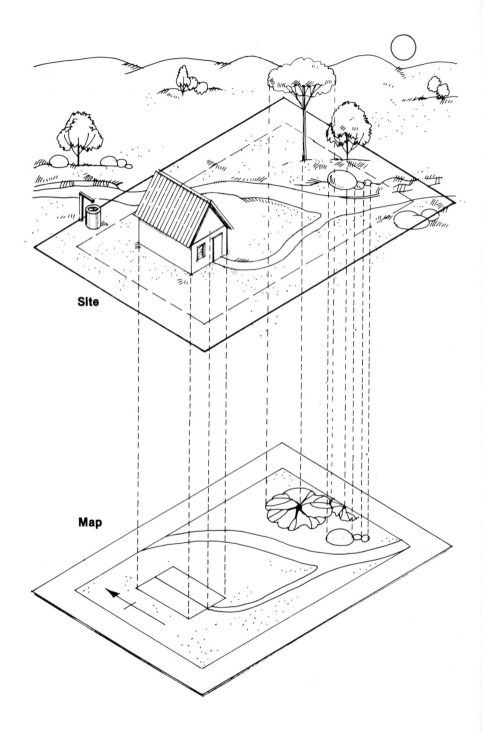

Site

Map

11 Purpose of this manual

In this manual, you will learn what you need to know about topography to help you choose a good site for your freshwater fish-farm, and to design and build fish ponds, reservoirs for storing water, and small dams. You will also learn how to draw your own topographical maps and how to use topographical maps that are already available.

To do all this you will learn:

- How to measure distances, angles, slopes, and height differences,
- How to set out straight lines, perpendiculars, and parallels in the field;
- How to determine horizontal and vertical lines;
- How to survey an area of land to find its size and its high or low, flat or sloping features (called the relief);
- How to make simple surveys that will help you when you are ready to build your fish-farm;
- How to prepare and how to use topographical plans and maps;
- How to calculate areas and volumes.

You will learn some of the technical language that land surveyors and civil engineers use. This way, you will be able to discuss your plans and projects with them more easily, and you will better understand books on topography, engineering and surveying.

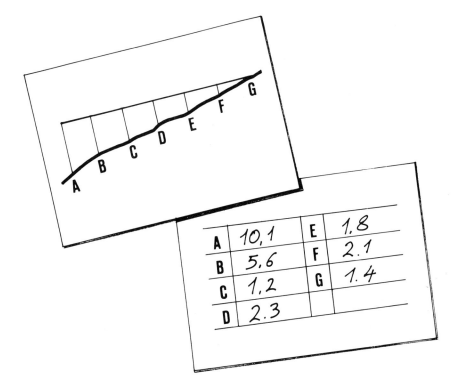

A	10,1	E	1,8
B	5,6	F	2,1
C	1,2	G	1,4
D	2,3		

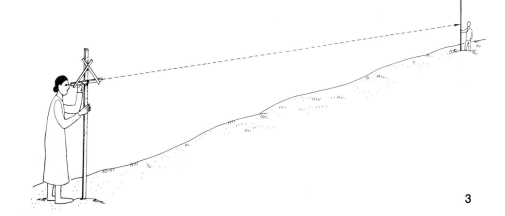

3

12 Why do you need to know about topography?

Choosing a site

1.   In other books in this series, **Simple Methods for Aquaculture**, *FAO Training Series* (4 and 6), you learned how to study water and soil on a site before deciding to build a reservoir for storing water and a fish-farm on it. **Topography is also very important when you are choosing a site**. Good fish-farm construction is possible only with the right topography.

2.   After you choose a possible area of land for your fish-farm, you will need to measure:

- its size;
- the slope of its ground surface;
- its elevation (height) in relation to the source of water you want to use.

You will also need to find out:

- the distance between the source of water and the location of the ponds;
- the best way to supply water to the ponds;
- the easiest way of draining the ponds.

3. You will need **to measure distances** in different ways. You may have to measure very long distances; to do this, you must know how to measure distances along a straight line, and how to keep this line straight. In addition, when you measure on a slope, you will need to find the horizontal distance rather than the distance on the ground.

4. When you are looking for **a site for a small dam**, you will find that the best choice is a narrowing valley where the stream slope is not too great and the valley walls are steep. You can use an existing topographical map to help you find such a valley, or you can measure a valley yourself to see if it is a good dam site.

5. When you have chosen the best site for your fish-farm, **topographical methods** will help you to decide whether you can do all the work that is needed on it.

6.   After you choose a possible site for your fish-farm, you will need **to survey the site**, keeping in mind your plan for the farm. For this survey you will measure distances, directions, areas, slopes and height differences in more detail.

7.   To do this, you will have to draw **a detailed topographical plan**. This plan will show the position of boundaries, the different heights of land forms such as hills, and the location of existing physical features such as paths, roads, streams, springs, forests, rocks and buildings. Such a plan is very important because it gives you the basic horizontal and vertical elements of the area, which will guide you in your design of the farm. It shows you the direction the water will have to take, from higher to lower points. It guides you in choosing where the water-supply canal, the ponds and the drainage ditches will go. It becomes the basis for estimating how much earth you will have to move as you build, and how much all the work will cost.

8.   All **the physical features of your fish-farm** depend directly on the topography of the site. These features include the type, number, size and shape of the fish-ponds, and how they are placed in relation to each other. The supply of water and the type of drainage also depend on the topography of the site.

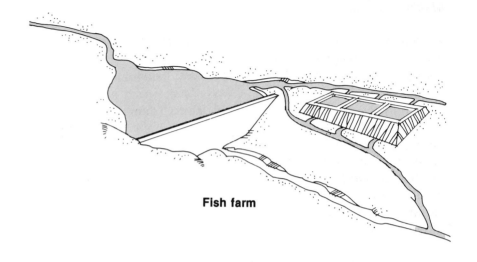

Fish farm

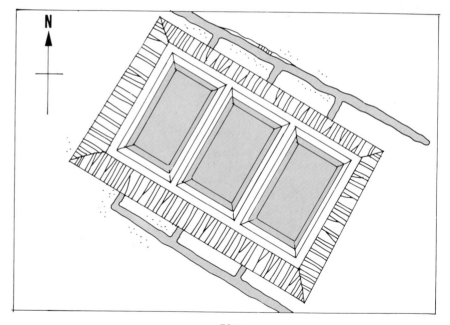

Plan

Making a construction survey

9.   Once you have made a detailed survey of the site you have chosen, and designed the fish-farm or dam (see next volumes in this series), you will use topographical methods to help guide you as you build.

10.   You will need to make sure that your fish-ponds regularly get the right amount of water. To do this, you must build a watersupply canal with the right size and bottom slope. First, you will need to stake out **the water supply canal**, along its centre-line. You will then need to tell the workers helping you exactly how wide, deep and long the canal must be, and how much earth they must remove at each point of the canal.

11.   You will need to stake out **the bottom area of each pond** and tell the workers how much earth to remove and where to transport it. This will allow you to drain all the water out of the ponds in a natural way. It will then be easy to harvest your fish and to manage your pond.

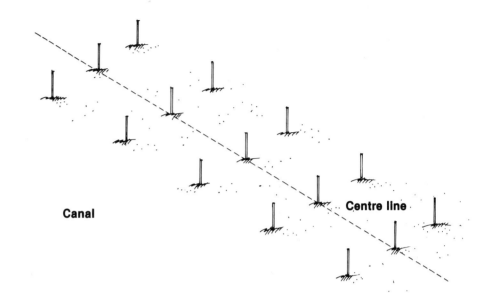

Canal   **Centre line**

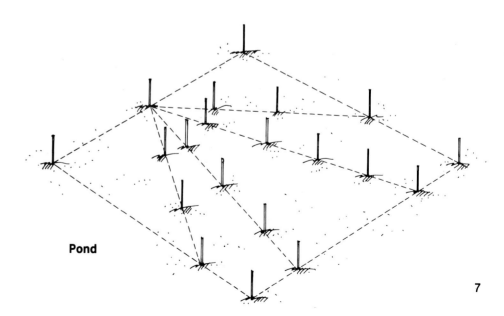

Pond

7

12. You will need to stake out **the dikes of each pond** and show the workers where to remove soil and where to add soil. You will also need to mark the location, height and width of each dike, as well as the slopes of their walls. Usually, you will need to set out **perpendicular*** (crossing) and **parallel*** (side-by-side) lines to do this.

13. You will need to follow the exact plan of your fish-farm as you work. To do this, you will have to be sure where each structure should be built, and you will have to check these locations during construction. You will need to measure differences in height between the different parts of the farm to make sure that the water will flow naturally in the right direction. The water will have to flow, for example, from the water source to the ponds, from the pond inlets to the outlets and from the outlets into a drainage ditch, which carries the water away from the farm site.

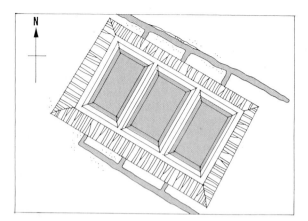

Plan

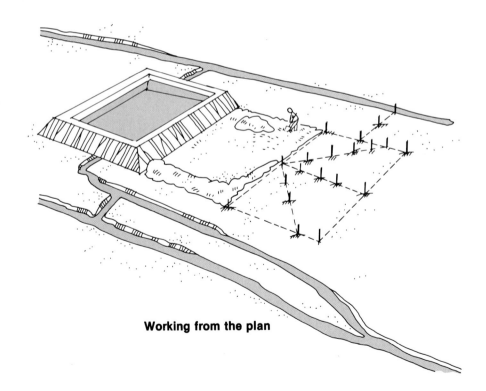

Working from the plan

8

14. In Water for Freshwater Fish Culture, *FAO Training Series* (4), you learned how to use simple topographical methods:

- for finding the surface area and the water volume of ponds (see Section 20, pp. 20-25);
- for finding the water flow of a stream (see Section 33, pp. 50-53);
- for using a weir (see Section 36, pp. 77-78);
- for measuring the heads of water pipes and siphons (see Sections 37 and 38, pp. 87, 91);
- for choosing the site of a dam (see Section 41, pp. 98-101);
- for estimating the volume of a reservoir (see Section 42, pp. 102-110).

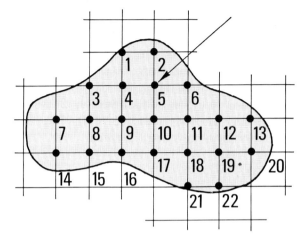

Measuring the volume of a pond

Choosing a site for a reservoir

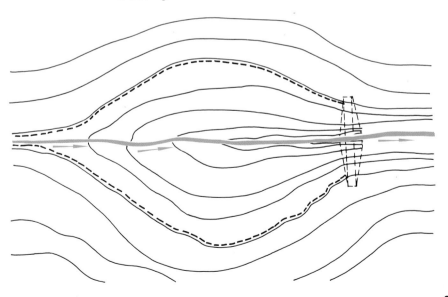

Studying your soils

15.   In Section 13 of **Soil and Freshwater Fish Culture**, *FAO Training Series* (6), you learned that the qualities of soils vary depending on the topography of the area. Shallow soil is found on sloping land and deep soil is found on flatter land, for example. You learned that alluvial soil, which is found in sedimentation plains, often contains large amounts of clay. The clay in this soil helps it to retain water, and makes it a good material for building dams.

16.   You will need to use topographical methods to draw a map showing the different kinds of soil present in an area of land. In **Soil and Freshwater Fish Culture**, Sections 24 and 25, you learned how to use two of these methods: reconnaissance surveys and detailed soil surveys.

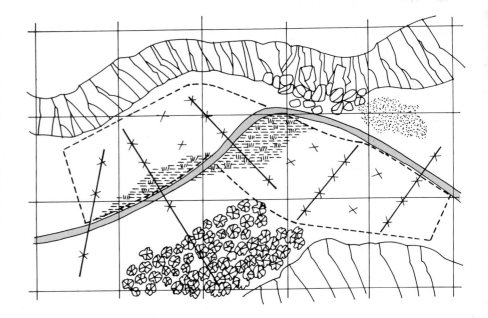

A map for studying your soil

13 There are two kinds of lines in topography

1. Almost all topographical methods are based on lines. There are two kinds of lines, lines of **measurement** and lines of **sight**.

- **Lines of measurement** may be either horizontal or vertical or they may follow the level of the ground. These lines are clearly plotted in the field with markers to show the exact path along which you will measure. A line of measurement can be:
    - **a straight line**, which runs in one direction between two marked end-points;
    - **a broken line**, which changes direction more than once between two marked end-points, with each point at which the direction changes also marked;
    - **a curved line**, which is marked like a broken line, but with markers much more closely spaced so that the curve is clearly followed.
- **A line of sight** is an imaginary line that begins at the eye of the surveyor and runs towards a fixed point. Lines of sight are either horizontal or oblique (between horizontal and vertical).

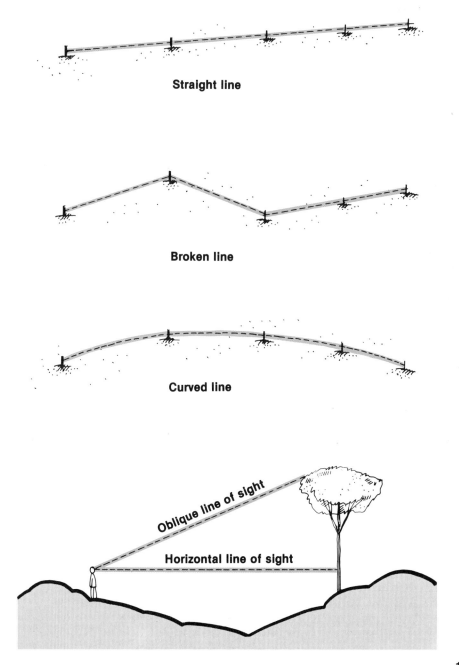

Straight line

Broken line

Curved line

Oblique line of sight

Horizontal line of sight

14 How to plot lines in the field

1. **Lines of measurement** are always plotted on the ground either as one **straight line** or as many **connected straight lines**. The markers that show where the line goes can be pegs, small concrete pillars, simple wooden stakes or ranging poles (see next Section).

2. **Lines of sight*** are always straight lines. The object or point you look toward, called the **point of reference**, is marked either by a ranging pole or a levelling staff (see Section 50).

3. **Vertical lines of measurement** can be formed with the help of a plumb-line (see Section 48).

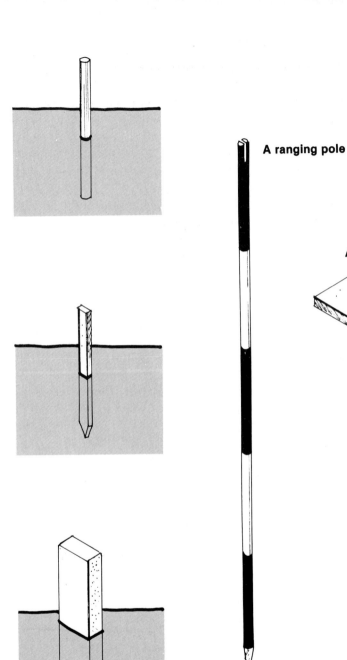

A ranging pole

A plumb-line

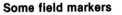

Some field markers

15   How to make and use markers to plot your line

1.   You can make **wooden pegs** to use when you plot your line. Get straight pieces of wood 3 to 8 cm in diameter and 0.1 to 1 m long. With a knife, shape the pieces at one end to make sharp points so that they can be easily driven into the ground. Your pegs will last longer if you make them out of hard wood and coat them with used engine oil to prevent rotting.

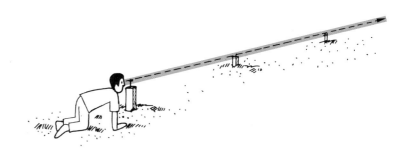

2.   You can use **iron pegs**, made of cut pieces of iron rod or tube about 1.3 to 2 cm in diameter. You can also use long wire nails. Iron pegs last longer than wooden pegs, but they cost more and they are heavier and more difficult to carry when you are working in the field.

3.   When you have a point on the ground that you will need to refer to for a long time, you can mark it with a small, **upright pillar** made of concrete. Such pillars should be from 15 to 30 cm square, and 10 to 60 cm high. You can build them on the site, placed on a small concrete base.

The nail shows the exact point

Note: to make your measurements in the field more accurate, you will often need to **mark a point** on your peg or pillar. This will show exactly where you must take the measurement, or place a measuring instrument. To do this, you can drive a nail into the flat top of the wooden peg, or you can set a nail into the top of the concrete pillar.

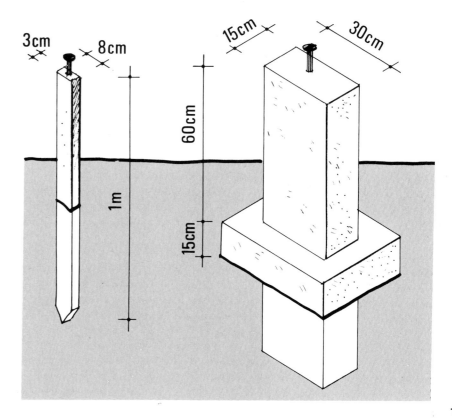

4. **Ranging poles** are the most commonly used markers in topographical surveys. **Ranging poles** are long, thin poles. You can use them to mark a point on the ground that you need to see from a distance. You can easily **make your own ranging pole**. Get a straight wooden pole, 2 to 3 m long and 3 to 4 cm thick. Shape the lower end into a point to make it easier to push into the ground. At the other end of the pole, cut a slit 5 cm deep into the top side. Then, starting from the top end, paint one 40 cm long section red; paint the next 40 cm section white. Continue painting the pole in alternating red and white sections until you reach the end.

5. Sometimes you will need **to sight a ranging pole from a long distance**. To make the pole easier to see, fasten two small flags of different colours, one above the other, near the top of the pole. Or you can place a 15 x 25 cm piece of white cardboard in the slit at the top of the pole.

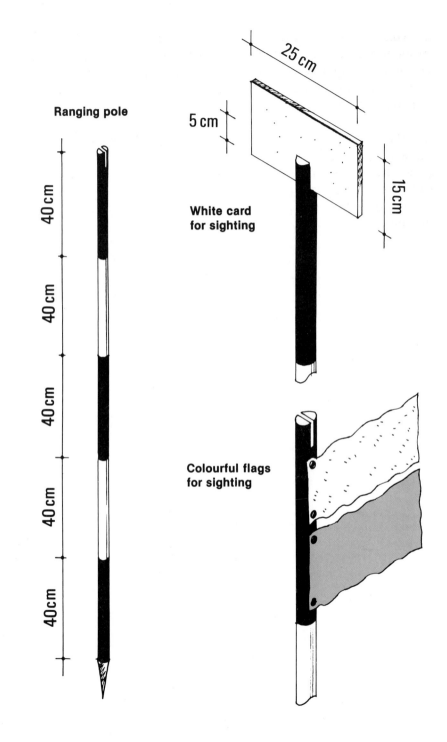

Ranging pole

40 cm

40 cm

40 cm

40 cm

40 cm

25 cm

5 cm

15 cm

White card for sighting

Colourful flags for sighting

6. You must always **drive ranging poles vertically** into the ground. To check that your pole is vertical, take a few steps back and look at it. If it seems straight, walk one-quarter of the way around the pole, and check that it also looks straight from the side. Adjust it if necessary until the front and side views are both vertical.

7. At times you will have to centre **a ranging pole over a marker** and leave it in position for some time. To do this, you can use a series of **guys**. These are ropes or wires which you will tie around the pole, and fasten to pegs in the ground. You can also use guys with the pole on **hard ground**, whenever you cannot drive the pole deep enough into the ground to keep it in a vertical position.

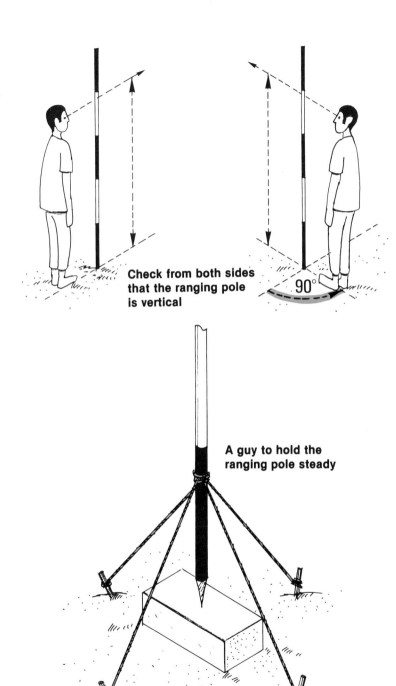

Check from both sides that the ranging pole is vertical

90°

A guy to hold the ranging pole steady

16 How to set out a straight line between two points

1. When you carry out a simple survey, you will often need to **set out straight lines between two given points**, called A and B, which are more than 50 m apart. To do this, you will **"range" line AB**. This means that you will plot **intermediate points** along line AB at intervals preferably shorter than 30 m.

2. When you range a line, you will face one of two possible situations:

- you can see point A from point B and vice versa;
- you cannot see point A from point B. In this case, an **obstacle** (a forest, river, lake, etc.) is in the way.

Setting out a straight line between two points visible from each other

3. You want to set out line AB. Mark the beginning of the line, point A, with a ranging pole. Then mark the end of the line, point B, with another ranging pole. You will now need an assistant to help you.

4. Stand about 1 m behind ranging pole A and look at ranging pole B. Your assistant should stand at ranging pole B. Ask him to walk, carrying another ranging pole, about 40 paces from B towards A, and stop.

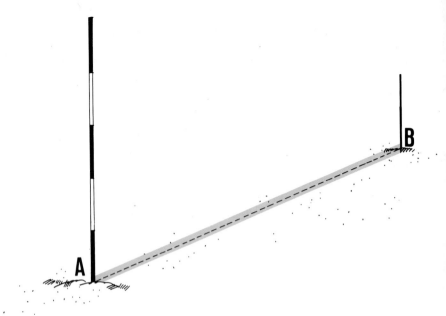

Setting out line AB with ranging poles

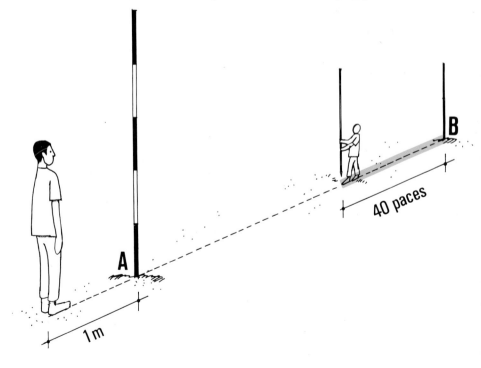

5. Ask your assistant to move slowly sideways while he holds the ranging pole vertically in one hand between his thumb and forefinger. When the ranging pole he is holding hides ranging pole **B**, ask him to stop and to drive his pole vertically into the ground. This is intermediate point C.

6. Ask your assistant to walk 40 paces toward you, from C toward A. Then repeat the same procedure as above with a fourth ranging pole. Mark the new intermediate point D.

7. If the distance from D to A is greater than 50 m, you should repeat the same procedure and mark the next intermediate points E, F, G

Note: always make sure that the ranging poles are **vertical**.

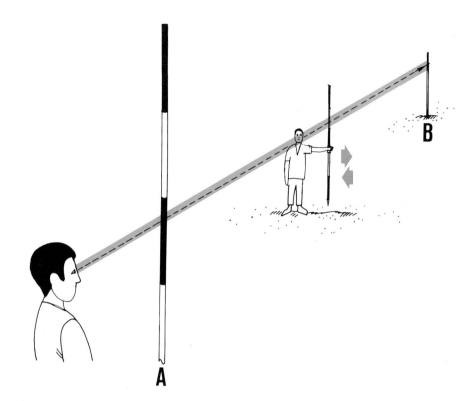

Make sure the ranging poles line up exactly

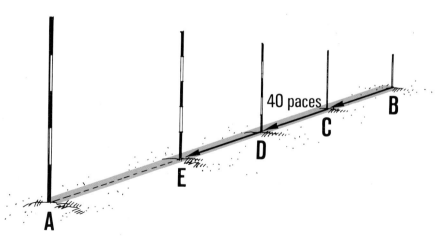

40 paces

Setting out a straight line between two points when you cannot see one from the other

8.   You have to set out line AB, which runs through a forest. Mark points A and B with ranging poles. Choose a point X, which is beyond point B, and which you can see clearly from point A. Mark point X with a ranging pole or a marking peg. Then set out a line as above from point A to point X, avoiding the forest.

9.   Look at Section 36 of this manual, and learn how to drop a perpendicular. Then, from point B, **drop perpendicular BC** onto line AX. The lines will cross at point C.

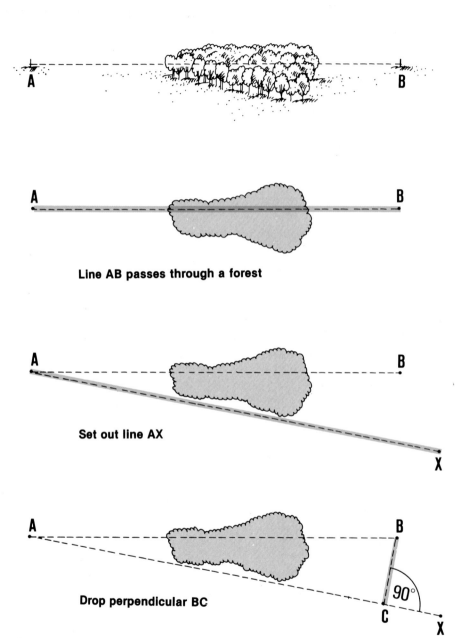

Line AB passes through a forest

Set out line AX

Drop perpendicular BC

10. Choose a point D on line AX, close to the forest, and **set out perpendicular DY**. Point Y must be on the **other side** of line AB.

11. Measure **horizontal distances** AD, AC, and CB, using one of the methods described in Chapter 2.

12. Intermediate point E will be the place where line DY intersects line AB. To find its exact location, you must calculate horizontal distance DE using the formula:

$$DE = AD \times (CB \div AC)$$

13. **To mark point E**, you must measure this distance DE horizontally. Starting from D, pace off the distance DE along line DY. Mark intermediate point E with a ranging pole.

14. Walk along line AX to the other side of the forest. **Set out a perpendicular** FZ close to the forest, using one of the methods described in Section 36. Point F is on line AX; point Z is **beyond** line AB.

15. Measure horizontal distance AF (see Chapter 2).

16. Point G will be the intersection of lines AB and FZ. To find it you must first calculate horizontal distance FG as:

$$FG = AF \times (CB \div AC)$$

17. Measure this distance FG horizontally. From F, measure along line FZ **to determine point G**. That is, the intersection of line FZ and line AB. Mark intermediate point G with a ranging pole.

18. You have now clearly laid out and marked line AB in the field as line AGEB.

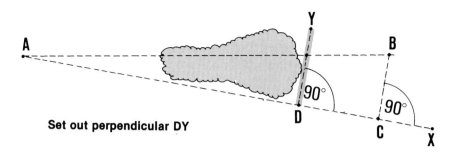

Set out perpendicular DY

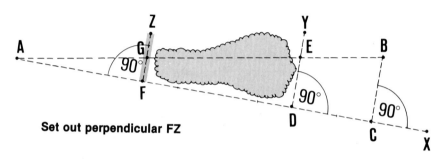

Set out perpendicular FZ

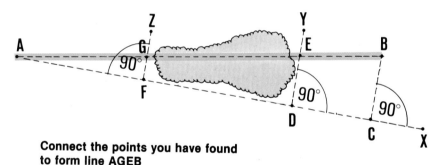

Connect the points you have found to form line AGEB

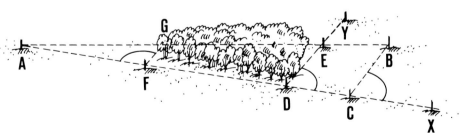

17 How to prolong a line you have marked in the field

1. You will often need to make a straight line you have marked longer; this is called **prolonging a line**. As in the previous section, you will have to consider two different situations:

- prolonging a line where there is no obstacle;
- prolonging a line behind an obstacle.

Prolonging a line where there is no obstacle

2. Mark a straight line AB in the field with a ranging pole at each end. **If you are working alone**, take a ranging pole and walk away from point B, following the direction of line AB as closely as you can. After you have walked about 40 paces, stop and turn around to face ranging poles B and A.

3. Hold your ranging pole vertically in front of you between your thumb and forefinger. Then move slightly sideways, if necessary, until your pole seems to **hide ranging poles B and A** from your view. Drive your pole into the ground in a vertical position.

4. **Step back** 1 to 2 m along the line and check to see if ranging poles B and A are still hidden behind your ranging pole. If they are not, move your pole a little to the left or right, and step back and check again. Repeat this procedure until your pole is in the right position. This then marks point C, which prolongs line AB.

5. **If you are working with an assistant**, stand 1 to 2 m behind ranging pole A to determine a **line of sight AB**. Your assistant should stand by ranging pole B.

6. Ask your assistant to walk, carrying a ranging pole, about 40 paces from ranging pole B in the direction away from you. He should then stop and turn around to face you.

7. While your assistant holds his ranging pole vertically, ask him to move to the left or right until ranging poles A and B hide his ranging pole. At that point, direct your assistant to drive his ranging pole vertically into the ground. This marks point C, which prolongs line AB.

20

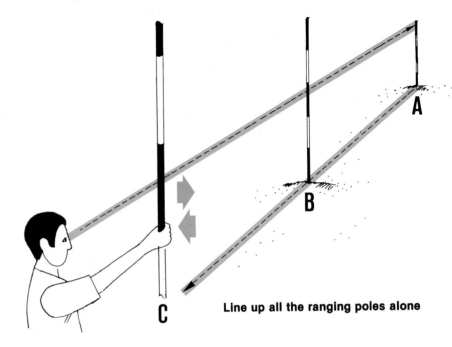

Line up all the ranging poles alone

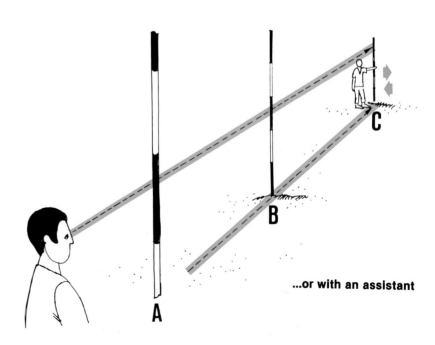

...or with an assistant

Prolonging a line behind an obstacle

8. You want to prolong line AB to a point behind a forest. Set out **perpendiculars AX and BY** from points A and B respectively, using one of the methods described in Section 36.

9. On these two perpendiculars, **measure equal horizontal distances** AA′ = BB′. You must make sure that this distance is far enough along the perpendiculars so that the line joining points A′ and B′, when prolonged, will just clear the obstacle.

10. Prolong line A′B′ through C′ and D′, **well past the obstacle**, using the method described above in steps 2-7.

You must prolong line AB through the forest

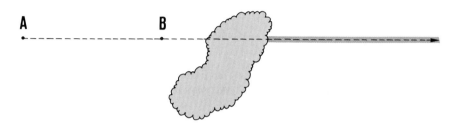

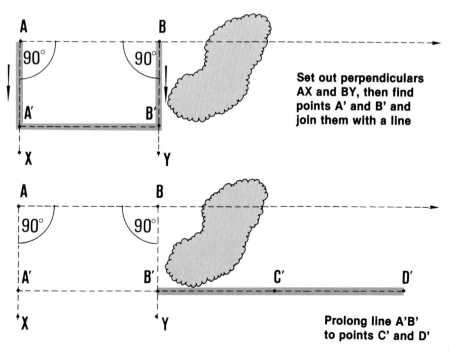

Set out perpendiculars AX and BY, then find points A' and B' and join them with a line

Prolong line A'B' to points C' and D'

11.   At points C' and D' set out **perpendicular lines** C'Z and D'W (see Section 36).

12.   On these two perpendiculars, measure **horizontal distances equal to AA'** (see step 9 above) and determine points C and D. Mark these points with ranging poles. You have now prolonged line AB with line CD.

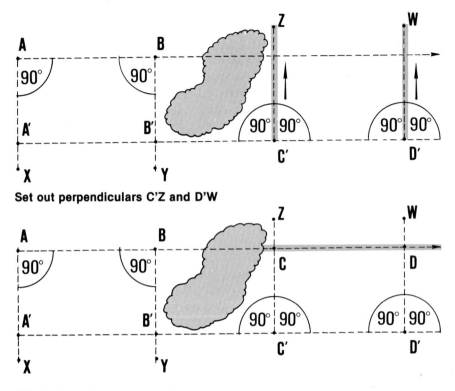

Set out perpendiculars C'Z and D'W

Find points C and D and join them to prolong AB

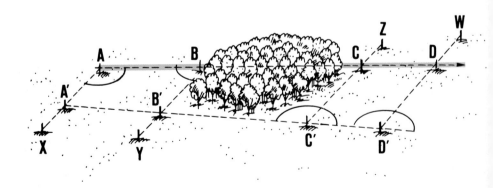

2 MEASURING HORIZONTAL DISTANCES

Measuring distances along straight lines

1. In topographical surveys, you measure distances along **straight lines**. These lines either join two fixed points or run in one direction starting from one fixed point. They are plotted in the field with pegs, pillars or ranging poles.

Expressing distances as horizontal measurements

2. You should always measure distances as **horizontal distances**. You may have to measure on ground which has no slope, or only a very small slope that is less than or equal to 5 percent (see Section 40). The distance measured on this type of ground will be equal to or very close to the horizontal distance. **When the slope of the terrain is greater than 5 percent**, however, you will have to find the horizontal distance. To do this, you must either correct any measurements you made along the ground (see Section 40) or use another method of measurement (see following sections). Unlevelled ground also requires particular methods of measurement.

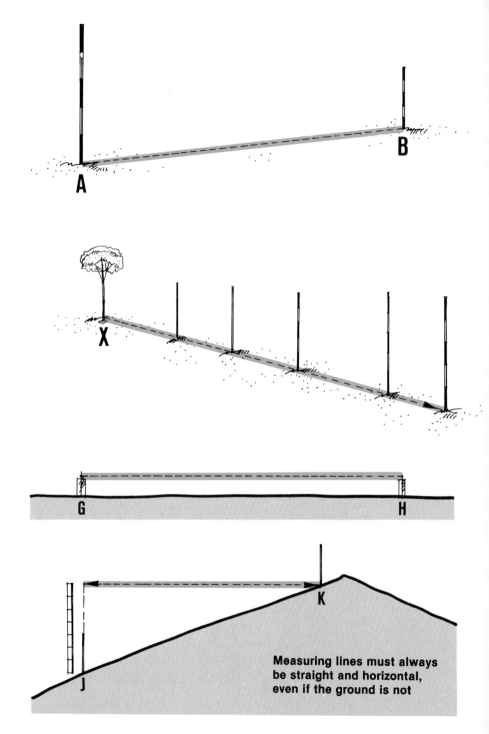

Measuring lines must always be straight and horizontal, even if the ground is not

Measuring distances along lines that run through obstacles

3. Usually, you will be able to reach all the points of the straight line you want to measure. But in some cases an obstacle such as a lake, a river or a cultivated field will be in the way, and you will have to take **indirect measurements**. This means that you will calculate horizontal distances along the original straight line (see Section 29).

Choosing the most suitable method

4. There are many good ways to measure distances. The method of measurement you use will depend on several factors:

* the accuracy of the result needed;
* the equipment you have available to use;
* the type of terrain you need to measure.

In the following sections, you will learn how to use the various methods of measurement. Table 1 will also help you to compare these methods and to select the one best suited to your needs.

Chaining with a rope

Calculating perimeter lengths

5.   The perimeter of an area is its outer boundary. The length of **the perimeter of regular geometrical figures** can be calculated from the mathematical formulas given in **Annex 1**, at the end of this manual.

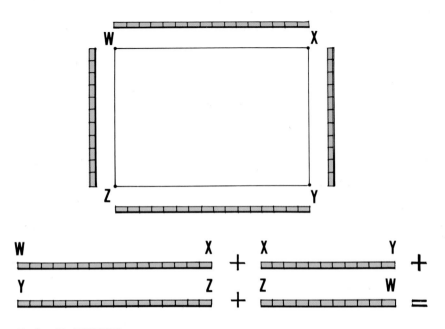

Perimeter WXYZW

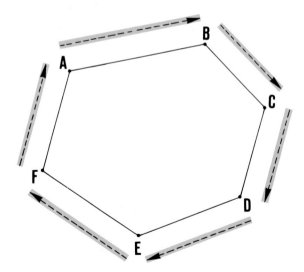

AB + BC + CD + DE + EF + FA = Perimeter ABCDEFA

TABLE 1
Distance measurement methods

Section [1]	Method	Distance	Error per 100 m [2]	Remarks	Equipment [3]
22 *	Pace count	Medium to long	1 to 2 m or more	For quick, rough estimates	None
22 *	Pacing with a passometer or podometer	Long	1 to 2 m or more	For quick, rough estimates	Passometer or podometer
21 *	Ruler	Short	0.05 to 0.10 m	Especially useful for sloping ground	*Ruler* (mason's level, plumb line)
23 **	Chaining using a rope	Medium to long	0.5 to 1 m	Cheap	Liana or *rope*, string, marking line
24 **	Chaining using a band or tape	Medium to long	less than 0.05 m	Best results with steel lines	Steel band, measuring tape
25 **	Chaining using a chain	Medium to long	0.02 to 0.10 m	Stronger quality	Surveying chain
27 ***	Clisimeter	Medium	1 to 2 m	For quick and rough estimates	Clisimeter (lyra-) (*2 m stadia staff*)
28 ***	Stadia	Medium to long	0.1 to 0.4 m	For quick and accurate measures	Telescope with stadia hairs, levelling staff

[1] * Simple   ** more difficult   *** most difficult.

[2] Error increases as the terrain becomes more difficult (slope, vegetation, obstacles).

[3] In addition to ranging poles (setting out the line), marking pegs (intermediate points), and notebook/pencil. *In italics*, equipment you can build yourself, as explained in text.

21   How to measure short distances with a ruler

1.   To measure short distances, use a measuring stick called a ruler, 4 to 5 m long. You can make your own by following the steps below. A ruler is particularly useful for measuring horizontal distances on sloping ground.

Making your own ruler

2.   Get a piece of wood which is straight and flat. It should be about 5 cm wide, and a little more than 4 m long. You can also use a shorter length, if the distance you are going to measure is shorter.

3.   It is best to use a planed piece of wood, but if you cannot get one you may use a straight wooden pole. If you use a pole, you should plane at least one of its surfaces.

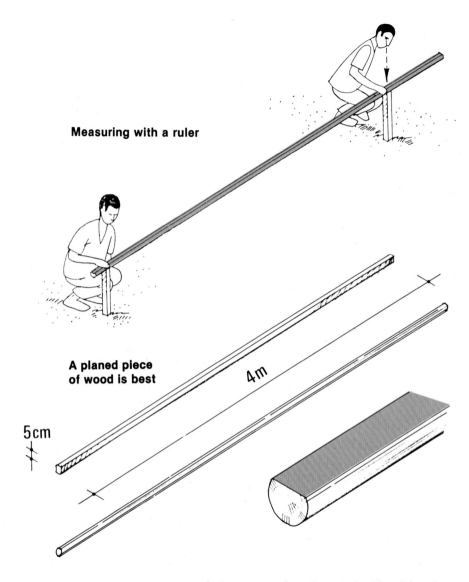

Measuring with a ruler

A planed piece of wood is best

5 cm

4 m

... but you can also plane one side of a pole

4. You should now add **graduations** to it. Graduations are marks which show exact measurements, in centimetres, decimetres, metres, etc. To do this, you need to get two **ready-made measuring tapes**, such as the 2 m ribbon tapes that tailors use. Glue one of these measuring tapes onto the planed face of your piece of wood. Take care to align the zero mark of this tape with one of the ends of the piece of wood. Glue the second tape next to the end of the first tape; this should reach near the end of the piece of wood. Drive several small nails through the tapes to secure them better.

5. You can also **make the graduations yourself**. Using a measuring ruler or tape, mark the graduations on the piece of wood with a pencil. With a knife or saw, make a straight, shallow cut every 10 cm. A carpenter, with his tools, can help you to do this more accurately. Identify the graduations with **numbers** (for example, every 0.5 m) using paint or ink; or you can use a piece of hot wire to burn the graduations and their numbers into the wood.

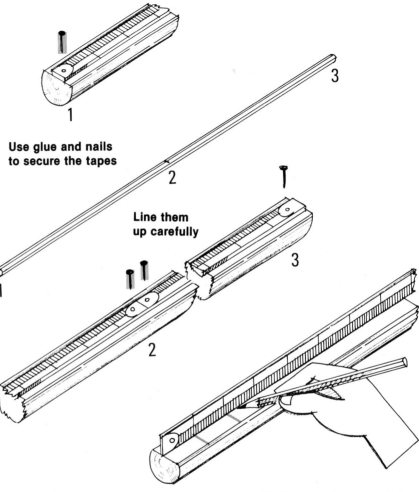

Use glue and nails to secure the tapes

Line them up carefully

Mark your ruler accurately

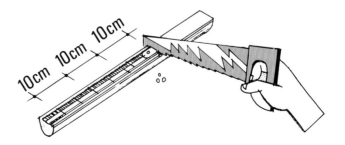

Measuring short distances with your ruler

6. When you have to measure a short distance on **horizontal terrain**, mark each end of the distance with ranging poles. Place your ruler on the ground with its end at the first ranging pole, making sure the ruler follows the straight line. Put a marking peg at the other end of the ruler. Then take the ruler and place its first end at this marking peg. Continue in this way until you reach near the end of the line, keeping an accurate count of the number of ruler lengths. You will usually need to use only part of the ruler's length to measure the last part of the line. Take care then to read the graduations on the ruler correctly.

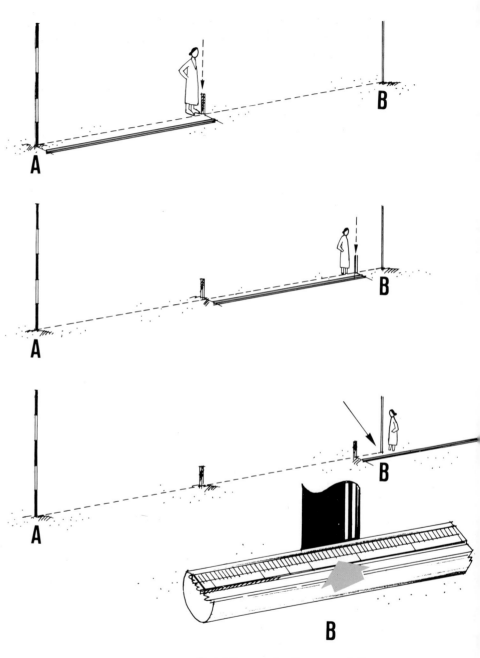

Read the graduations carefully

7.   When you need to measure a distance **on sloping terrain**, your ruler will be very useful for finding **horizontal distances**. You proceed downhill, and for each measurement:

● make sure that the ruler is horizontal, using a **mason's level** (see Section 61);
● determine the point where you need to place the marking peg, using a **plumb-line** at the end of the ruler (see Section 48).

Note: when you measure a distance on sloping ground, remember that you should proceed **downhill**.

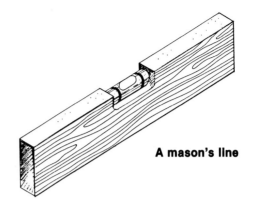

A mason's line

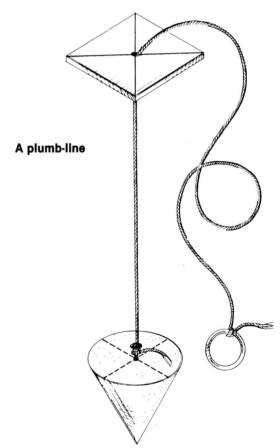

A plumb-line

22 How to measure distances by pacing

1. You may measure distances roughly by **pacing**. This means you count the number of normal steps which will cover the distance between two points along **a straight line**. Pacing is particularly useful in reconnaissance surveys, for contouring using the grid method (see Section 83) and for quickly checking chaining measurements (see Sections 23 to 25).

2. To be accurate, you should know **the average length of your step when you walk normally**. This length is called **your normal pace**. Always measure your pace from the toes of the foot behind to the toes of the foot in front.

Count your steps as you walk

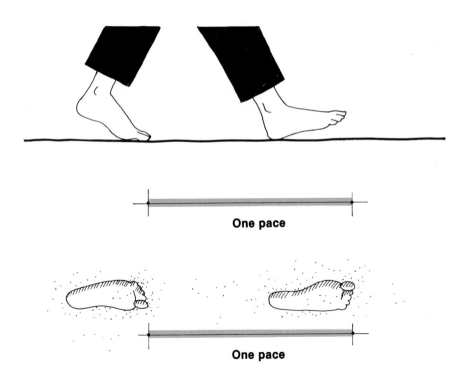

One pace

One pace

Finding your own pace factor

3. To measure the average length of your **normal pace** (the pace factor, or **PF**):

- take 100 normal steps on horizontal ground, starting with the toes of your back foot from a well-marked point, A, and walking along a straight line.
- mark the end of your last step with peg B, at the toes of your front foot.
- measure the distance AB (in metres) with, for example, a tape and calculate your pace factor PF (in metres) as follows:

$$PF = AB \div 100$$

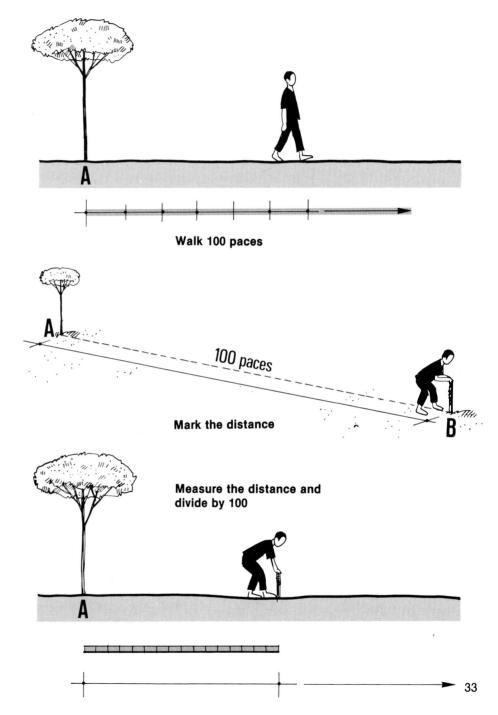

Walk 100 paces

100 paces

Mark the distance

Measure the distance and divide by 100

Example

If for 100 paces, you measure a distance of 76 m, then your pace factor is calculated like this: PF = 76 m ÷ 100 = 0.76 m.

Note: to determine a more accurate pace factor:

- walk over a **longer distance** (at least 250 paces);
- **repeat** the measurements at least three times and calculate **the average PF**.

Example

For 250 paces, you measure successively 185 m, 190 m and 188 m; in total, for 3 × 250 = 750 paces, you have walked 185 + 190 + 188 m = 563 m; your average pace factor, PF = 563 m ÷ 750 = 0.75 m.

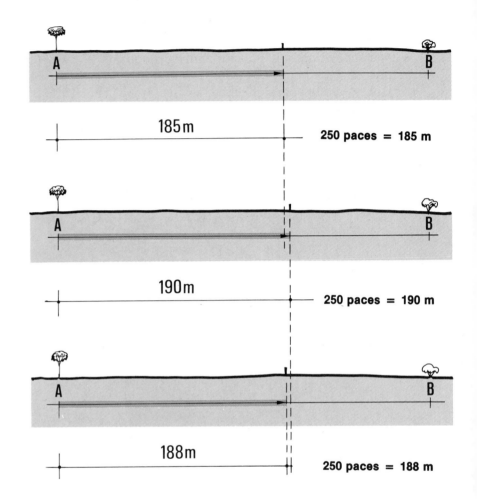

185 m   250 paces = 185 m

190 m   250 paces = 190 m

188 m   250 paces = 188 m

4. Your pace factor will vary, depending on **the type of terrain** you are measuring. Remember that:

- your pace will be shorter in tall vegetation than in short vegetation;
- your pace will be shorter walking uphill than walking downhill;
- your pace will be shorter walking on sloping ground than on flat ground;
- your pace will be shorter walking on soft ground than on hard ground.

To get the best results, you should first make your paces as nearly the same length as possible. To do this, walk over **known distances**, both on level ground and on uneven or sloping ground. Adjust your pace so that it is as regular as possible.

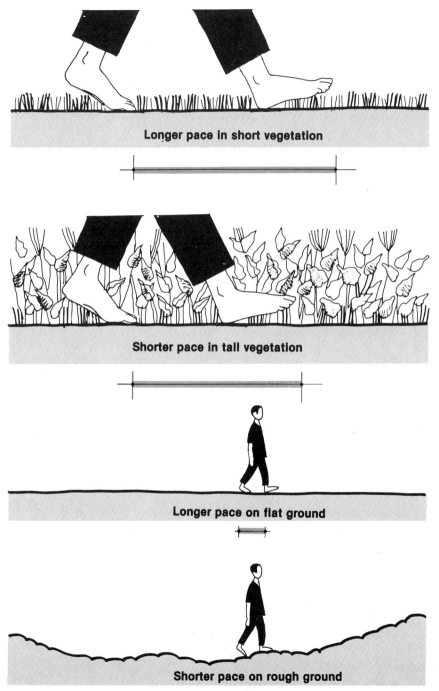

Longer pace in short vegetation

Shorter pace in tall vegetation

Longer pace on flat ground

Shorter pace on rough ground

35

Measuring horizontal distances by pacing

5.   Clearly plot the straight lines you have to measure, using wooden pegs or ranging poles. If necessary, remove any high vegetation standing in the way.

6.   Walk along the straight lines, carefully counting your steps.

7.   Multiply the number of steps N by your pace factor PF (in metres) to get a rough estimate of the distance in metres, as follows:

> **Distance (m) = N × PF**

Example

To measure ABCD, pace distances AB = 127 steps; BC = 214 steps; and CD = 83 steps. ABCD = 127 + 214 + 83 = 424 steps. If PF = 0.75 m, ABCD = 424 × 0.75 m = 318 m

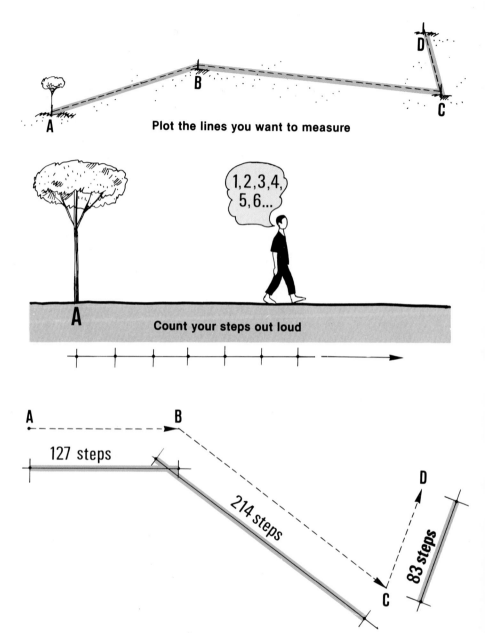

Plot the lines you want to measure

1,2,3,4, 5,6...

Count your steps out loud

A   B

127 steps

214 steps

D

83 steps

C

Note: to avoid errors when counting your steps:

- count only double steps or **strides**, and multiply the total count by 2;
- take count of the **hundreds** with your fingers (using one finger for each hundred steps);
- take count of the **thousands** by ticking them off on paper;
- when **crossing** obstacles such as fences and small streams, estimate the number of steps, strides or half-steps it would take to cross them.

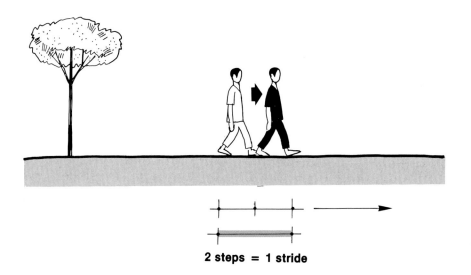

2 steps = 1 stride

1 step

Estimate the size of obstacles

Pacing with a passometer or a pedometer

8.   You can register your paces mechanically by using a simple device called a **passometer**. The passometer is about the size of a watch. You should wear it on a point near the centre of your body, attached to a belt or waistband for example. At each pace you take, the jolt of your step makes a pointer in the passometer turn. This pointer shows the number of paces.

9.   **The pedometer** is a similar device, but it registers distance. This is usually expressed in kilometres and fractions of kilometres.

10.   You should check these two devices for accuracy before you use them. To check a passometer, walk a few hundred paces, counting them carefully. Then compare your total count of paces with the number of registered paces, and adjust the device as necessary. To check a pedometer, walk at a **normal** pace along a straight line over a known distance. Compare this distance with the registered distance, and adjust the device as necessary.

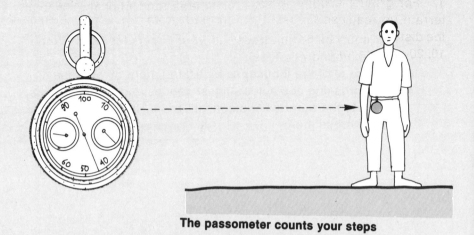

The passometer counts your steps

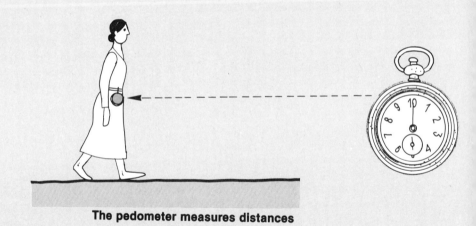

The pedometer measures distances

23 How to chain with a rope

1. For greater accuracy in measurement, especially over difficult terrain, you can use **a measuring line** made from rope. Depending on the distances you need to measure, you can make a measuring rope 10, 20 or 30 m long.

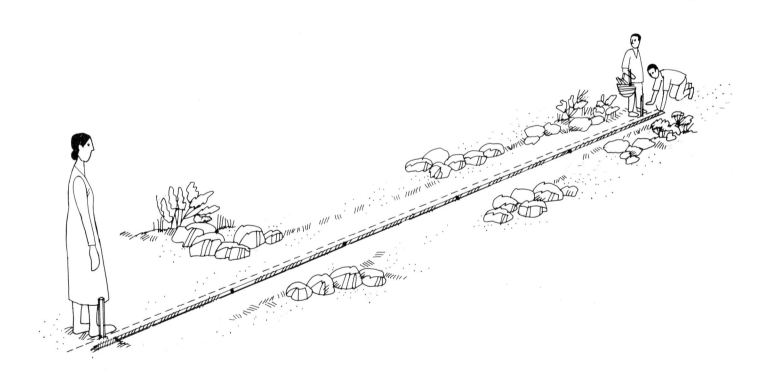

Making your own measuring rope

2.   Get a rope 1 to 1.5 cm thick, made of **natural fibres**. Ropes of man-made fibres, such as nylon, may change over time, but natural materials, such as jute, will shrink or stretch very little. A piece of used sisal rope is better to use than a new one. You can also use a piece of supple liana, which you can easily find in the forest.

3.   Put the first mark – the zero **mark** – about 20 cm from one end of the rope. From this point, accurately measure the length you need one metre at a time. Leave about 20 cm at the other end of the rope. Mark each metre point with durable **waterproof paint**, dye, ink or coloured wax. Keep these metre marks as thin as possible to avoid inaccurate measurements. You can use thin string for the marks instead, threading the string through the rope so that it does not shift position.

4.   Reinforce **the two ends** of your measuring rope. To do this, tightly wind some light string around the last 10 cm of each end of the rope.

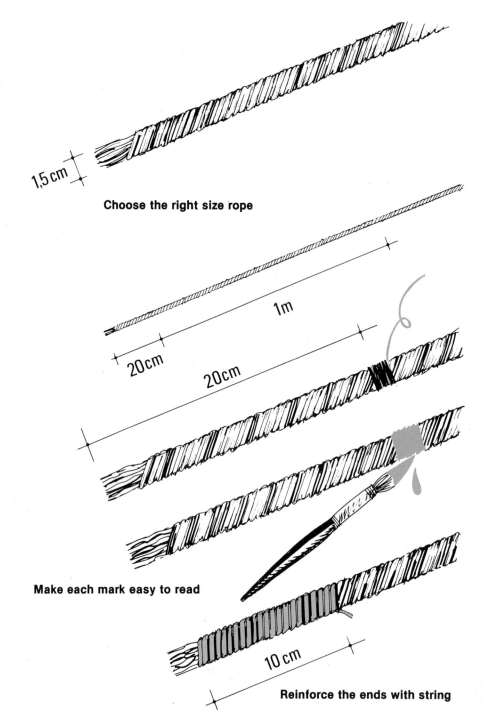

1,5 cm

Choose the right size rope

1m

20cm

20cm

Make each mark easy to read

10 cm

Reinforce the ends with string

40

Measuring horizontal distances with a rope

5. Clearly plot the straight lines you want to measure, using wooden pegs, for example. On either side of each of these lines, clear a narrow strip of ground completely, removing vegetation and large stones.

6. If the distances are shorter than your rope, or about the same length, you can take their measurements directly. To do this, carefully stretch the rope from one peg to the next. If a distance falls between the metre marks on your rope, measure this shorter length with a ruler or a tape graduated in centimetres.

7. If the distances are longer than your rope, you will need to use one of the chaining methods described later (see Section 26). These methods can be used with all measuring lines, including ropes, bands, tapes or chains.

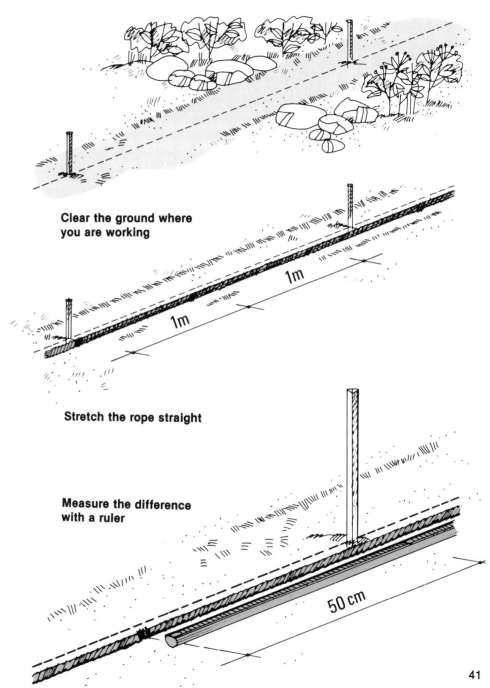

Clear the ground where you are working

Stretch the rope straight

Measure the difference with a ruler

41

24 How to chain with bands or tapes

1. You can buy bands and tapes in stores. **A measuring band** is made of a strip of steel, usually 6 mm wide and 30 or 50 m long. Metres, decimetres and centimetres are clearly marked on the band. Bands are wound onto an open frame, with a spindle and handle for rewinding.

2. **Measuring tapes** are made of steel, metallic cloth or fibreglass material. They come in lengths of 10 to 30 m or more. They are usually marked at 1 m intervals, with the first and last metres graduated in decimetres and centimetres. They are wound into a case, with a handle for rewinding. Tapes can present some problems. Steel tapes can easily become twisted and break. Cloth tapes are less precise than the others, since they often vary slightly in length.

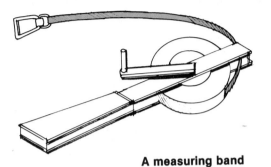

A measuring band

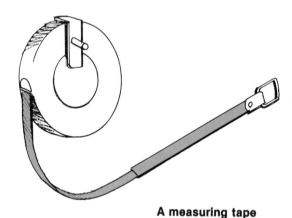

A measuring tape

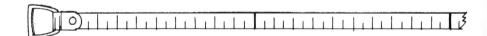

Measuring horizontal distances with a steel band or a tape

3. Plot the straight lines you need to measure. If the lines are the same length as your measuring band or tape or shorter, you can measure the distances directly. To do this, stretch the band or tape from one peg to the next one.

4. If the lines are longer than your band or tape, use one of the methods described later (see Section 26).

Note: you should pull bands and tapes tight, so that they do not sag, especially when you are measuring long distances. But, you should avoid over-stretching them (especially fibreglass tapes), since this could lead to errors.

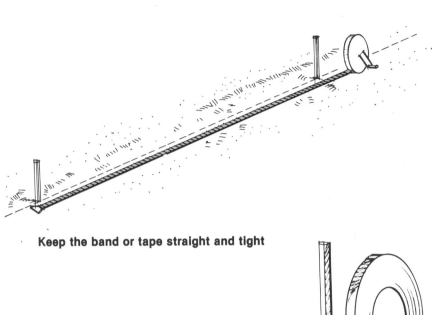

Keep the band or tape straight and tight

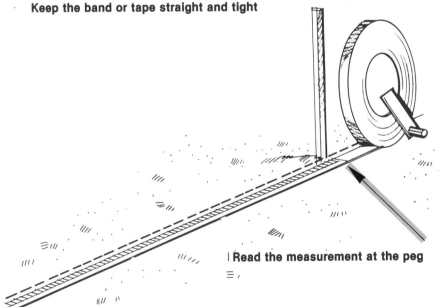

Read the measurement at the peg

25 How to chain with a surveyor's chain

1. **Surveyor's** chains are also sold in stores. They are made of a series of steel links; each link is the same length, usually 20 cm. The links are attached to each other by steel rings. **The length of one link** includes its straight portion, its two rounded ends, and the two half-rings that connect it to the links on either side. Each metre of the chain is usually marked by a **brass ring**. At each end of the chain, there is a metal **handle** which you should include in the measurements. The total length of the chain is usually 10 or 20 m. Chains are less accurate than bands and tapes, but they are much stronger.

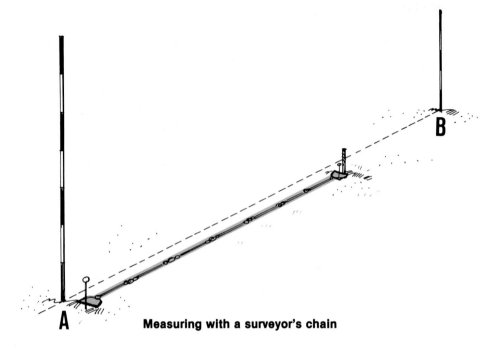

Measuring with a surveyor's chain

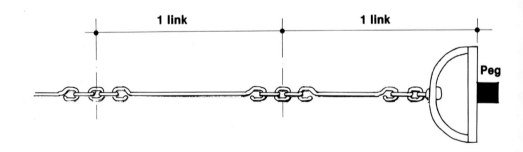

2. When you use a surveyor's chain, you should be careful of the following:

- make sure that the rounded end of one length does not remain on top of the one next to it. This can make the chain shorter. At the start of each survey, check for this by sliding the entire length of the chain through your hand and straightening all the links;
- avoid leaving the chain in the sun since the heat may cause the chain to become longer;
- pull the chain tight enough for accurate measurement.

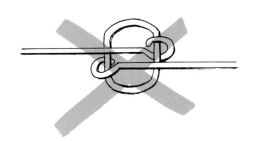

3. When using a chain for the first time, you should carefully measure **the length of each link**, using a ruler. Remember that this length includes the straight part and its two rounded ends, as well as the two connecting half-rings. At each end of the chain, the handle, one shorter link, and half the connecting ring make up the length of a link. After measuring the length of the links, check that 1 m of chain equals the expected number of links.

Make sure the links are straight

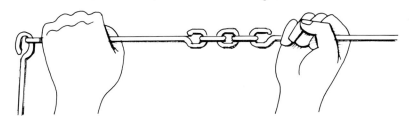

Example

If each link is 0.20 m long, there should be five links per metre of chain.

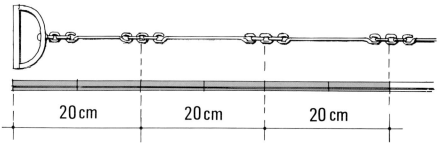

20 cm   20 cm   20 cm

Measure the length of each link with a ruler

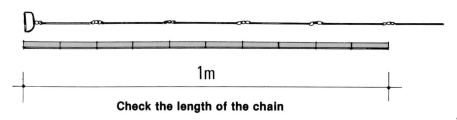

1m

Check the length of the chain

4. You should always fold the surveyor's chain as follows:

- take the two handles together in your left hand, doubling the chain;
- collect two links at a time with your right hand, putting them slantwise.

5. To unfold a surveyor's chain, hold the two handles in your left hand and throw the chain in the direction of the measurement you want to make.

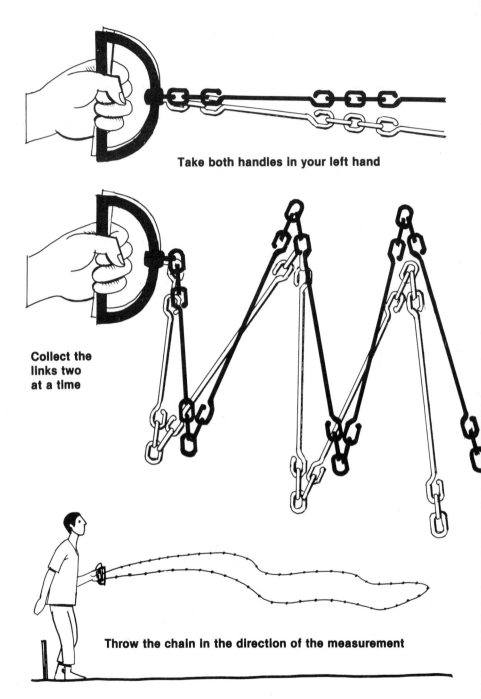

Take both handles in your left hand

Collect the links two at a time

Throw the chain in the direction of the measurement

Measuring horizontal distances with a chain

6.   The chain is used for measuring the lengths of straight lines, which should be marked at each end with a ranging pole. You will need an assistant to help you. The method of chaining you use depends on the type of terrain you are measuring (see Section 26).

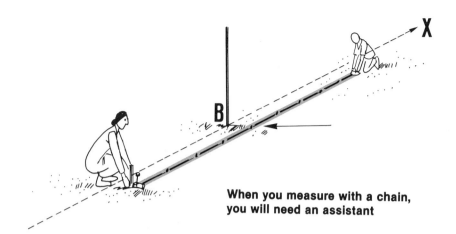

When you measure with a chain, you will need an assistant

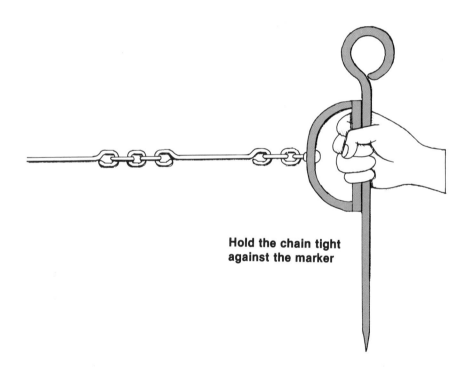

Hold the chain tight against the marker

26 How to measure distances by chaining

1.   As you have learned, measuring lines can be ropes, bands, tapes or surveyor's chains. When you measure long distances, the way you use the measuring line will depend on the slope of the terrain. When **the terrain is flat** or nearly flat (that is, with a slope of 5 percent or less -- see Section 40), you can measure the horizontal distances by following the ground surface. This method is usually used in measuring fish culture sites, where steeper slopes must be avoided. When the slope of **the terrain is steeper** than 5 percent, you should be especially careful when you measure the horizontal distances because in this case the surface measurement is always greater than the horizontal measurement.

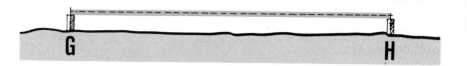

On flat ground, measure directly

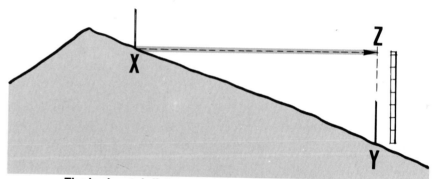

The horizontal distance is the true measurement

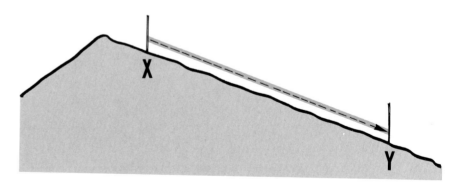

... the surface measurement is longer

Chaining over horizontal ground

2. Mark each straight line you need to measure with a ranging pole at each end. On lines longer than 50 m, place intermediate markers at regular intervals.

3. To measure long distances accurately, you will need **marking pins**. You can use thin wooden stakes about 25 cm long, which you can easily carry in a small basket. These marking pins will be driven vertically into the ground as you proceed with the chaining.

4. Chaining is carried out by two persons, a **rear chainman** and a **head chainman**. The **rear chainman** is responsible for the measurements. He notes the results. He also guides the **head chainman** to make sure that the consecutive measurements are made exactly along straight lines between the marked ground points.

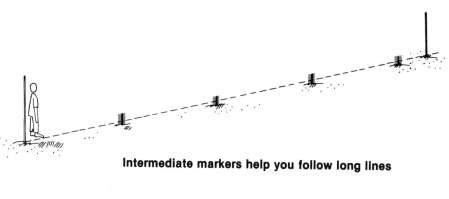

Intermediate markers help you follow long lines

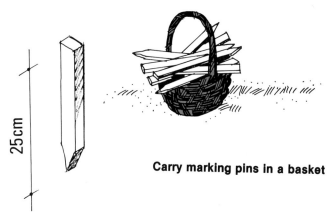

25 cm

Carry marking pins in a basket

The rear chainman guides the measuring and keeps the notes, while the head chainman takes the readings

5. Start the measurements at one end of the straight line. Remove the ranging pole and drive the first **marking pin** into the ground at exactly the same point.

6. The rear chainman places his end of the measuring line against this marking pin. The head chainman, taking with him a number of **marking pins**, walks away along the straight line with the other end of the measuring line.

7. The head chainman stops when the measuring line is stretched out tightly to its full length on the ground. He then looks towards the rear chainman. If the measuring line is not placed exactly along the straight line, the rear chainman then tells the head chainman how **to correct** the position of the measuring line.

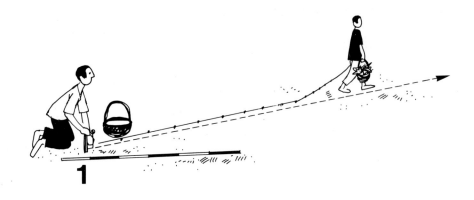

The rear chainman stays at the first point

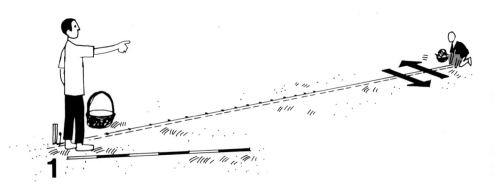

... and helps the head chainman find the second point

50

8. When the measuring line is correctly placed, the rear chainman signals to the head chainman to **place a second marking pin** at the end of the measuring line.

9. The rear chainman immediately **notes down this measurement**.

10. The rear chainman then removes the first marking pin, putting it in his basket, and **replaces the ranging pole** at the starting point.

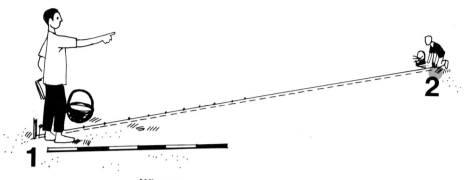

When the second pin is placed

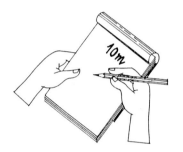

...the rear chainman notes the measurement

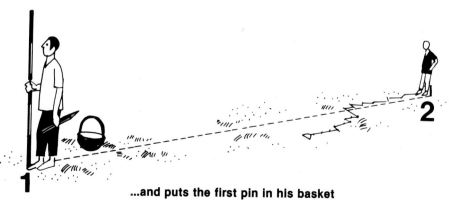

...and puts the first pin in his basket

51

11. Still holding their ends of the measuring line, both chainmen move forward along the straight line, **always keeping the measuring line well stretched**. This is particularly important when using a surveyor's chain.

12. **The rear chainman** stops at the second marking pin and places his end of the measuring line against it.

13. **The head chainman** tightens the measuring line along the ground, corrects its position following any directions from the rear chainman, and places a **third marking pin** at the end of the measuring line when signalled to do so.

14. The rear chainman **notes down this measurement**. Then he puts the second marking pin in his basket before moving on.

15. The process in steps 10 to 14 should be repeated along each section of the straight line until the end is reached.

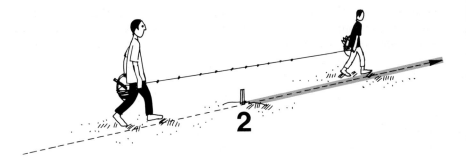

Both chainmen move forward

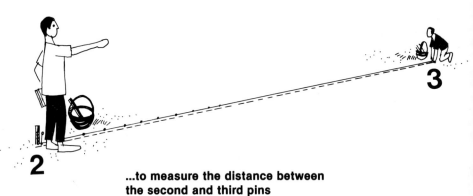

...to measure the distance between the second and third pins

The head chainman notes each distance

Note 1: when the end of the line is reached, the number of marking pins in the basket of the rear chainman shows **the number of complete measuring-line lengths measured**. You can use this to check on the measurements noted down.

Note 2: using a **set of 11 marking pins** makes it easier to keep track of the number of measurements completed. When the rear chainman has ten pins in his basket, ten complete measuring-line lengths have been measured. He **notes this down** and gives the ten pins back to the head chainman, **leaving the eleventh pin in the ground**; this is the starting point of a new series of measurements.

Example

Using a chain 10 m long, **the rear chainman** has marked 4 × 10 pins in his notebook. He has 6 marking pins in his basket. At the marking pin still in the ground, he has measured a distance of (4 × 10) + 6 = 46 chain lengths or 46 × 10 m = 460 m.

The number of pins in
the basket helps you keep count

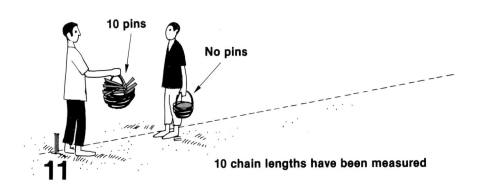

10 pins

No pins

11

10 chain lengths have been measured

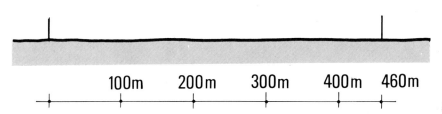

100m   200m   300m   400m 460m

Chaining over sloping ground

16.   When you are measuring on ground with a slope **greater than 5 percent** (see Section 40), you will need to use the measuring line differently to find the **horizontal distances**.

17.   Proceed as described in the previous section. Mark the straight lines with ranging poles at each end and intermediate pegs. Remember to **work downhill** for greater accuracy.

18.   The head chainman should hold the measuring line **horizontal**, above the ground, in this case.

19.   When the measuring line is in the right place and is fully stretched, the head chainman finds the exact point on which to place the marking pin, using a **plumb-line** (see Section 48).

20.   Keep proceeding in this way along the slope.

Note: **on steep slopes**, use a shorter measuring line (such as 5 m, rather than 10 m).

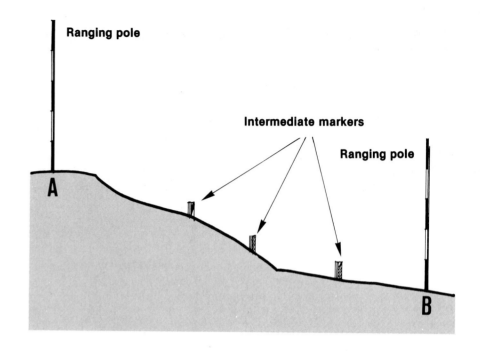

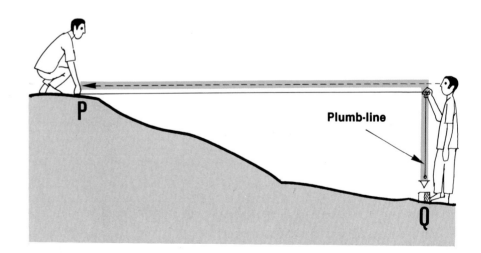

54

21.   As you measure on sloping ground, remember these important requirements:

- **horizontal** measuring line;
- **well-stretched** measuring line;
- **exact** placement of the **marking pins**.

Note: you may also measure along the ground on a slope. But to obtain horizontal distances, you will need to correct these ground measurements afterwards by using **mathematical formulas** (see Sections 40 and 50).

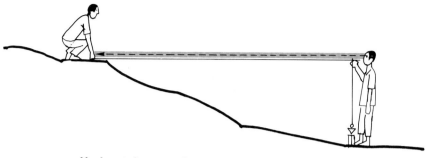

Horizontal measuring line

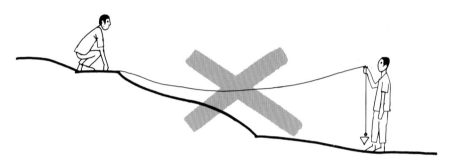

Well-stretched measuring line

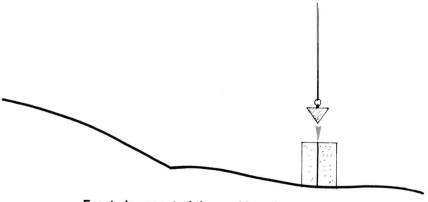

Exact placement of the marking pins

Chaining over irregular ground

22.   You may need to measure distances over irregular ground that has ridges, mounds, rocks, trenches or streams in the way. In such cases, you need to **lift the measuring line** above the obstacle. Make sure that you do the following:

- keep the measuring line **well stretched**. The head chainman may shorten it by looping it in his hand if necessary;
- keep the line **horizontal**, using a mason's level for the best accuracy (see Section 61);
- lift the back end of the measuring line **exactly above the marking pin**, using a **plumb-line** if necessary (see Section 48).

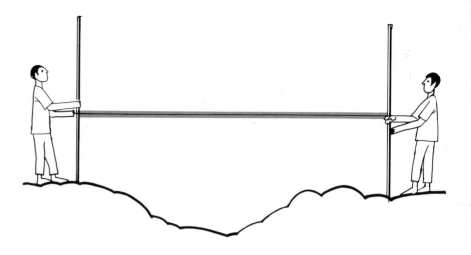

Lift the line over obstacles

Wrap it around your hand to shorten it

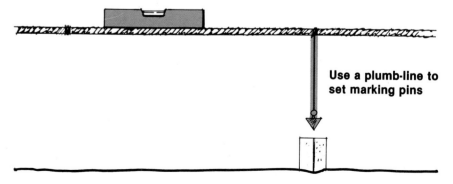

Level with a mason's level

Use a plumb-line to set marking pins

23. Instead of using a plumb-line, you can use **longer marking pins**, such as **ranging poles**, set vertically in the ground.

24. **In very hard or rocky soils**, you will not be able to use marking pins. In such cases, mark the points with objects you can see easily, such as painted rocks or blocks of wood. Make sure that your markers will not blow or roll away. Or, you can make a mark on the ground with a stick, or make a mark on a rock with chalk.

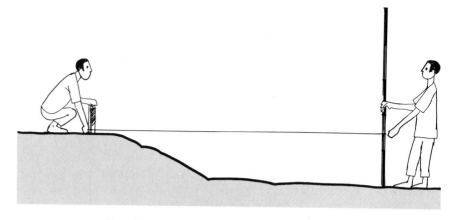

Ranging poles are taller than marking pegs

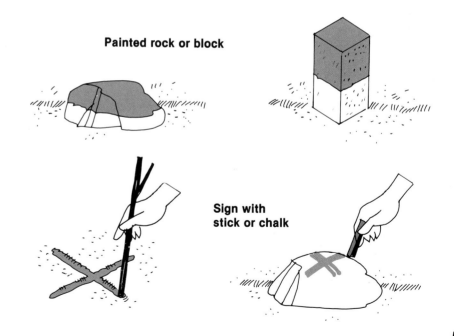

Painted rock or block

Sign with stick or chalk

Improving the accuracy of your chaining

25. To make your chaining more accurate, you should **repeat the measurements at least once**; start measuring at the point where you finished, and continue back along the line. This second measurement should not differ too much from the first one (see Chart 1).

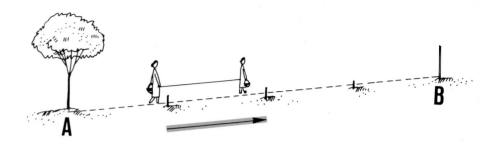

26. **If the two measurements agree**, you can calculate their **average value**. The average value is taken as the true measured distance.

Maximum permissible difference between two consecutive distance measurements, per 100 m

Steel band or tape	0.1 m
Other tape	0.2 m
Surveyor's chain	0.2 m
Home-made rope	1.0 m

Measure twice for accuracy

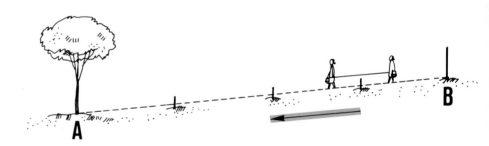

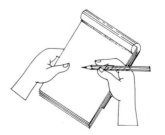

$$\frac{AB + BA}{2} = \text{Average}$$

Example

Using a surveyor's chain, you take the following measurements:

- first measurement: 312.6 m;
- second measurement: 313.2 m;
- real difference: 313.2 m - 312.6 m = 0.6 m;
- acceptable difference: 0.2 m × (312.6 m ÷ 100) = 0.2 m × 3.12 m = 0.62 m which is larger than the real difference and therefore agrees;
- average distance: (312.6 m + 313.2 m) ÷ 2 = 312.9 m

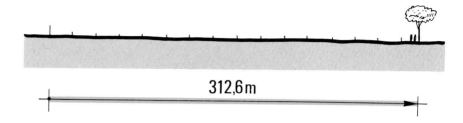

312,6 m

First measurement

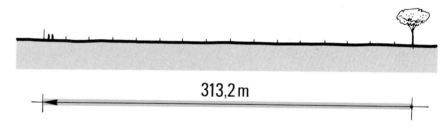

313,2 m

Second measurement

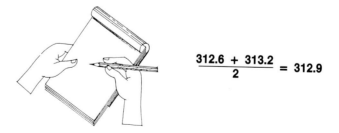

$$\frac{312.6 + 313.2}{2} = 312.9$$

27. If the two measurements differ by too much, you should take a third measurement. Compare this with the first two measurements. Then calculate the average value from the two most similar values, as shown above.

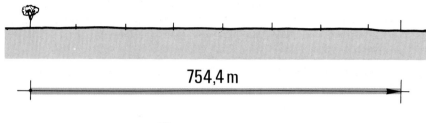

First measurement

Example

Chaining with a steel tape, you take the following measurements:

- first measurement: 754.4 m;
- second measurement: 753.2 m;
- real difference: 754.4 m – 753.2 m = 1.2 m;
- acceptable difference: 0.1 m × 7.54 m = 0.75 m, which is smaller than the real difference and therefore does not agree;
- third measurement: 753.9 m;
- difference 754.4 – 753.9 m = 0.5 m which is acceptable, being smaller than 0.75 m;
- average distance: (754.4 m + 753.9 m) ÷ 2 = 754.15 m.

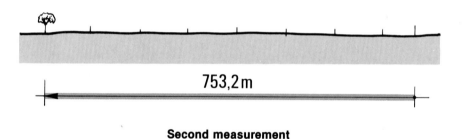

Second measurement

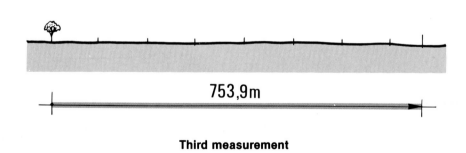

Third measurement

28. If you find you have very different measurements of the same line, you may not have been measuring along the **true straight line**. To reduce such errors, put more ranging poles on the line between the endpoints. If you tie white or brightly coloured pieces of cloth to the poles, you will be able to see them better. Also, be sure to guide the head chainman carefully as you measure.

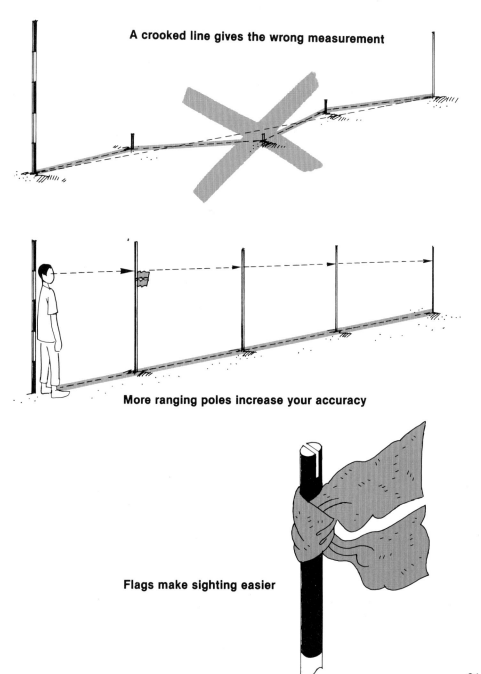

A crooked line gives the wrong measurement

More ranging poles increase your accuracy

Flags make sighting easier

29. Further improve the accuracy of your measurements by:

● inspecting the **full length** of the measuring line before using it to measure a series of straight lines;
● keeping a **uniform** tension on the measuring line during each measurement;
● accurately **marking** each point of measurement;
● keeping an accurate **count** of these points;
● using the right device, such as a ruler, to measure **distances less than the measuring-line length**, and knowing how to read the graduations on it (see Section 21).

Note: it is better if the head chainman holds the zero end of the measuring line. The rear chainman can then directly make and note down any intermediate readings.

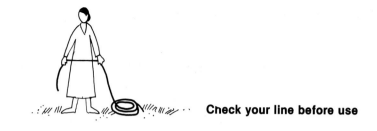

Check your line before use

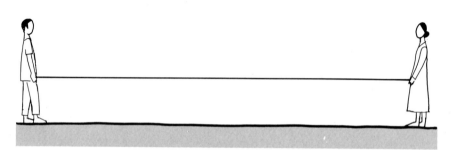

Always stretch it tight

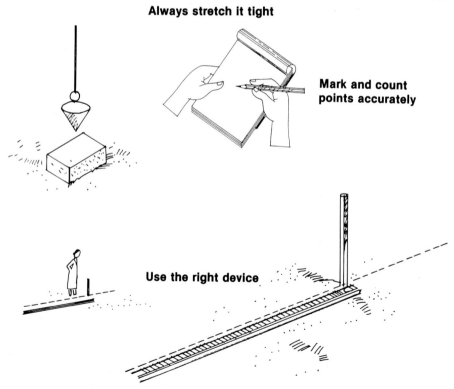

Mark and count points accurately

Use the right device

27 How to measure distances with a clisimeter

1. **The lyra clisimeter** is a simple instrument which can be used for measuring distances. It is also used for measuring ground slope (see Section 45). The clisimeter is not as accurate as a measuring line, but you can get a **quick estimate** of distances from it without having to walk the length of the line. The longer the distance you measure with it, however, the less accurate the measurement will be. For good estimates, **do not exceed 30 m distances. For rough estimates, you may measure distances up to 150 m**.

2. The lyra clisimeter consists of a **sighting device**, **a hanging ring**, and a **bottom weight** to keep the instrument in a stable vertical position. The clisimeter folds into this weight so that it can be easily carried.

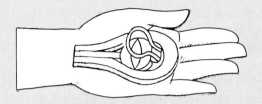

A folded clisimeter

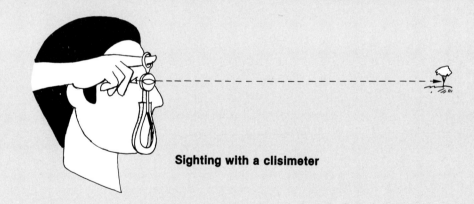

Sighting with a clisimeter

3. When you look through the sighting device, you will see **three vertical scales.** A scale is a series of marks along a line at regular intervals. You will use the **central scale**, the **stadimetric scale**, for measuring horizontal distances. Note that this central scale is made up of two parts:

- **The top part**, marked 150, 100 . . . 7 m;
- **The bottom part**, marked 150, 100 . . . 10 m.

4. To measure a distance with the clisimeter, you need **an assistant** to help you, and **a reference height** (called the **base**). The method you use with the clisimeter will depend on the kind of **base** you choose.

- You may use a **2 m base**, clearly marked on a wooden stake called a **stadia staff**. In this case, you will use the **top part** of the distance scale, marked BASE 2.00 m.
- Or you may take **the height of your assistant** as the base; in this case, you will use the **bottom part** of the distance scale, marked BASE 1.70 m.

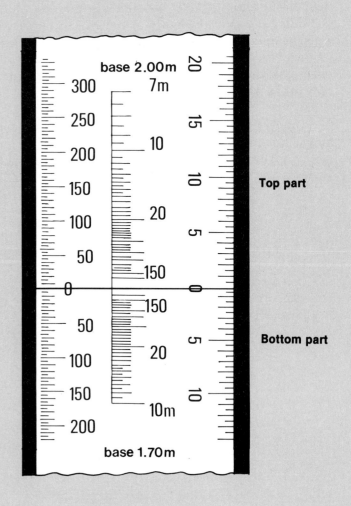

The scale inside a clisimeter

Making your own stadia staff

5. Get a straight piece of wood 2.50 m long. A rectangular stake with a cross-section of 8 x 4 cm is best, but you can use a round pole with a cross-section of 6 to 8 cm instead.

6. Get two wooden boards measuring 30 × 40 cm each.

7. Nail these boards along their centre lines 10 cm from each end of the stake, as shown in the figure.

8. Draw a horizontal line across the middle of one of these boards. This is called the **median line**.

9. From this line, **measure exactly 2 m along the stake**. You should reach a point near the middle of the second board. At this point draw a horizontal line across the board.

10. Using a pencil and ruler, divide the length of the stake between the two boards, which should be 1.70 m, into 10 cm sections.

11. **Paint** the two sections of the boards lying **outside the 2 m length in bright red.** Then paint, in red, the first 10 cm section next to each board and each alternate section in between.

12. Paint all the other sections of the boards and stake **in white**, including the 10 cm end sections of the stake. Your stadia staff is now ready to use for measuring distances.

Note: for short distances, you can use a **simpler staff**; get a pole or staff exactly 2 m long and paint it alternately in red and white, as described above.

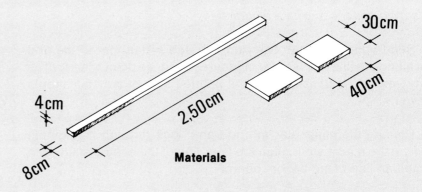

Materials

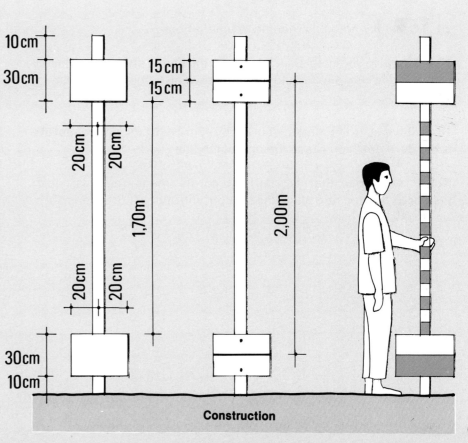

Construction

Measuring distances with a stadia staff

13. Send your assistant, carrying the stadia staff, out to the first point along the line you want to measure. There he places the staff as **nearly vertical as possible** and the painted side of the staff should face you.

14. Holding the clisimeter in one hand, look through its sighting device at the stadia staff. **Align the zero line** of the central scale with the median line of the bottom board.

15. Look at the **top part** of the central scale (BASE 2.00 m) of your clisimeter, and read the distance in metres at the graduation which lines up with the **median line of the top board**.

16. Carefully note this reading in a field-book.

17. Signal to your assistant to remove the stadia staff and replace it with a   marking pin. He should then move on to the next point to be measured.

18. Move up to the marking pin left by your assistant, and **repeat the procedure** until you reach the end of the line.

Note: remember that for fairly accurate measurements, each distance you measure along the straight line should not exceed 30 m.

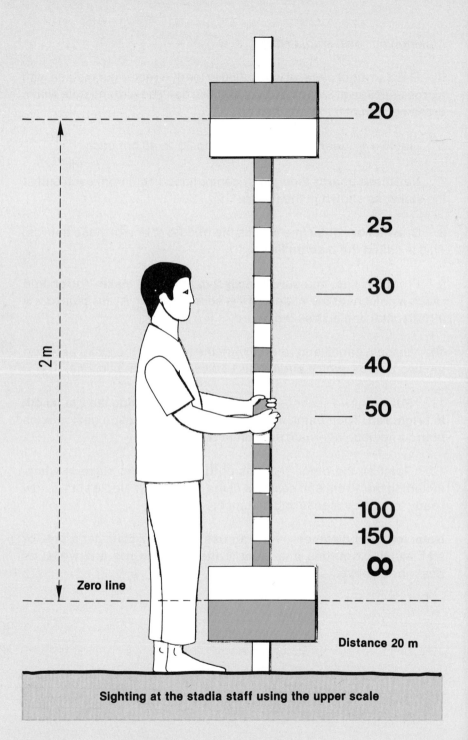

Sighting at the stadia staff using the upper scale

Measuring distances without a stadia staff

19.   If you do not have a stadia staff, you can use the **height of your assistant** as a reference instead. The height you need for this method is 1.70 m. Measure your assistant's height. If this differs much from 1.70 m, do one of the following:

- if your assistant is taller than 1.70 m, measure the height of his eyes or mouth from the ground and choose the height nearest 1.70 m;
- if your assistant is shorter than 1.70 m, ask him to place an object on his head (such as a can, a bottle or a block of wood) which will increase his height up to 1.70 m.

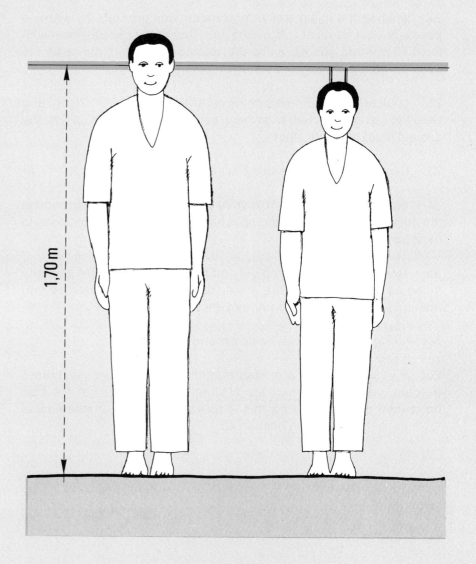

Find a reference point 1.70 m high

67

20.   Send your assistant out along the line you want to measure and, at the selected point, ask him to stand as straight as possible, facing you.

21.   Holding the clisimeter in one hand, look through the sighting device at your assistant. Align the zero line of the central scale with the 1.70 m mark you have chosen, such as the top of his head, his eyes or the top of a bottle carried on his head.

22.   Look at the bottom part of the central scale (BASE 1.70 m), and read the distance in metres at the graduation which lines up with the ground level under his feet.

23.   Carefully note this reading in your field-book.

24.   Signal to your assistant to drive a marking pin into the ground at the point where he was standing, and to move on to the next point to be measured.

25.   Move up to the marking pin and repeat the procedure as many times as necessary. For the greatest accuracy, each distance you measure along the line should not exceed 30 m.

Measuring distances on sloping ground

26.   If you are taking a measurement on a slope greater than 5 percent, you must correct the clisimeter reading to get the **true horizontal distance**. To do this, you need to use a mathematical formula, as explained in Section 40.

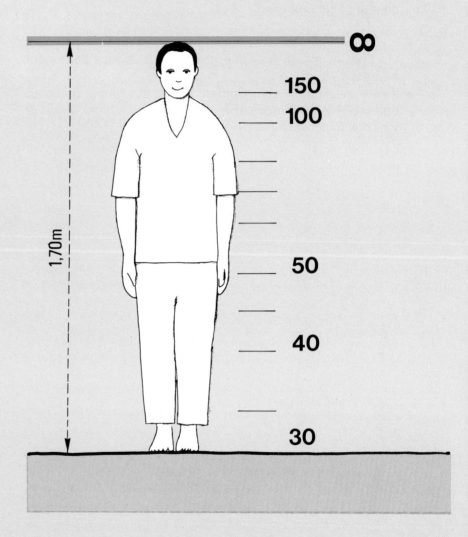

Sighting at your assistant using the lower scale

28   How to measure distances by the stadia method

The stadia method is rapid and accurate for measuring long distances, but to apply this method, you need to get **expensive surveying equipment** and learn how to use it. Therefore, only a brief description of the method is given here so that you can understand its basic principles.

1. The equipment used with this method includes a highly technical sighting device called **a telescope**. To use it, you must sight through two crossed hairs; there are also two extra horizontal hairs called **stadia hairs**. Most surveyor's levels (see Section 59) have these stadia hairs at an equal distance above and below the horizontal cross-hair.

2. To measure a distance, you will also need a **levelling staff** which is clearly graduated in centimetres (see Section 50).

3. Set up the surveyor's level at the point from which you will measure the distance. Signal to your assistant to place the levelling staff vertically at the next point of the line. The distance between you and the staff may be several hundred metres.

4. Look through the telescope and read the graduations (in metres) on the levelling staff that line up with the **upper stadia hair** and the **lower stadia hair**. Note these measurements down in your field-book.

5. Subtract the smaller reading from the larger reading. This represents the interval between the two hairs, called the **stadia interval**.

6. To find the distance (in metres), multiply the stadia interval by a fixed value called the **stadia factor**. It is given for each telescope, but on most instruments this factor equals 100.

Note: if you are working on sloping ground, you must correct this figure to find the true horizontal distance (see Section 40).

Example

- Upper stadia hair reading: 1.62 m;
- Lower stadia hair reading: 0.52 m;
- Stadia interval = 1.62 m - 0.52 m = 1.10 m;
- Stadia factor = 100;
- Distance AB = 1.10 m × 100 = 110 m.

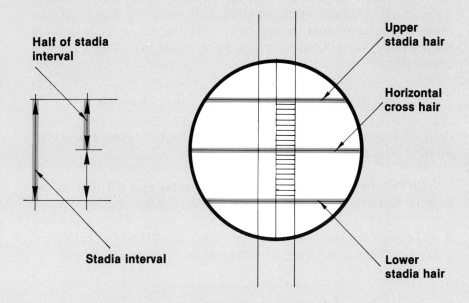

The scale inside a surveyor's telescope

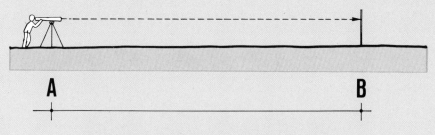

Sighting at a levelling staff with the telescope

29 How to measure distances that run through obstacles

1. To use the preceding methods, you must be able to walk over the whole length of each straight line and take direct measurements. Sometimes, however, there is an obstacle on the line that makes measuring the distance directly impossible. Such a line could be across a body of water such as a lake, a lagoon or a river, or across agricultural fields with standing crops. In these cases, you must take **indirect measurements** of a segment of the line. You will use some of the methods you learned in Section 16 for setting out a line across an obstacle.

Measuring a distance across a lake or an agricultural field

2. From point A on the line XY running through the obstacle, set out another straight line AZ, avoiding the obstacle.

3. On this new line, lay out a **perpendicular line CB** joining the original line at point B behind the obstacle (see Section 36).

4. Measure the two new line sections AC and CB and calculate the unknown distance AB from a mathematical formula as follows:

$$AB = \sqrt{AC^2 + BC^2}$$

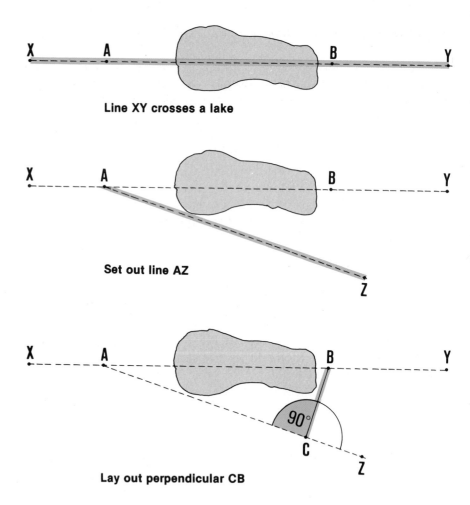

Line XY crosses a lake

Set out line AZ

Lay out perpendicular CB

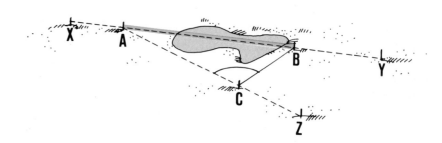

Measuring a distance across a river

5.   Here, the obstacle (a river) cannot be avoided, but you can see the points you need to measure from both sides of the river. There are several methods, based on geometry, which can be used. Two simple ones are described here.

6.   You need to measure distance GH across a river. Using ranging poles, prolong line GH back to point C. At G and C, lay out perpendiculars GZ and CX. On each of these lines, set out a point, E and F, so that they lie on **a straight line FY passing through H**, on the opposite bank. Measure accessible distances GE, GC and CF. Calculate the inaccessible distance GH as:

$$GH = (GE \times GC) \div (CF - GE)$$

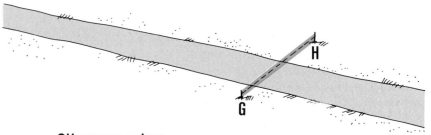

GH crosses a river

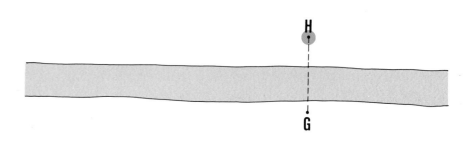

You can see point H from point G

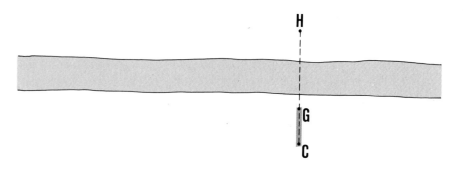

Prolong the line back to point C

71

Example

You wish to measure GH, across a river:

- prolong line GH back to C;
- lay out perpendiculars GZ and CX;
- select points F and E on line FEH;
- measure distances GE = 34 m; GC = 36 m; CF = 54 m;
- calculate GH = (34 m × 36 m) -: (54 m − 34 m); GH = 1 224 m ÷ 20 m = 61.2 m.

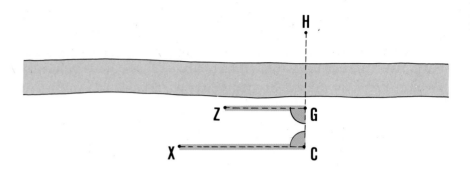

Lay out perpendiculars GZ and CX

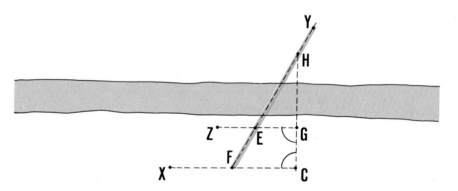

Find points E and F on line FEHY

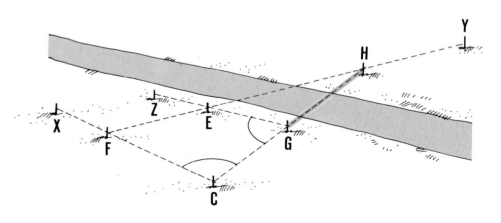

Calculate distance GH

7. You need to measure distance AB across a river. Lay out line BX perpendicular to AB on one river bank. Determine the point C of this perpendicular from which you will be able to sight point A across the river, **using a 45-degree angle** (see, for example, Section 36, step 63). Measure distance CB, which is equal to inaccessible distance AB.

Example

You need to measure distance AB:

- from B, lay out perpendicular BX;
- determine C, so that angle BCA = 45 degrees;
- measure BC = 67 m;
- distance AB = BC = 67 m.

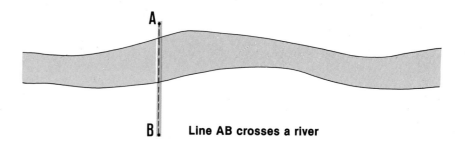

Line AB crosses a river

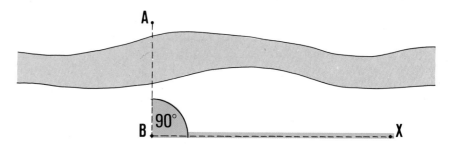

Lay out perpendicular BX

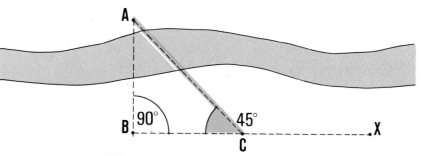

Lay out 45° angle BCA

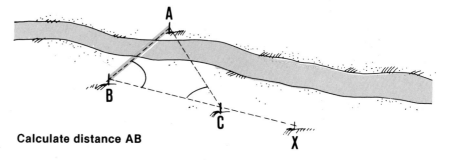

Calculate distance AB

3 MEASURING HORIZONTAL ANGLES

What is a horizontal angle?

1.   In topography, the angle made by two ground lines is measured **horizontally**, and is called a horizontal angle. You may replace these ground lines by **two** lines of sight AB and AC. These lines of sight are directed from your eyes, which form the **summit A** of the **angle BAC**, towards permanent landmarks such as a rock, a tree, a termite mound, a telephone pole or the corner of a building.

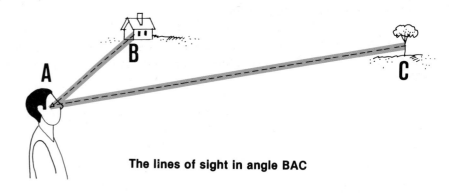

The lines of sight in angle BAC

Expressing horizontal angles

2. Horizontal angles are usually expressed in **degrees**. A full circle is divided into **360 degrees**, abbreviated as **360°. Note from the figure these two particular values:**

- a **90° angle**, called a right angle, is made of two **perpendicular** lines. The corners of a square are all right angles;
- a **180° angle** is made by prolonging a line. In fact, it is the same as a line.

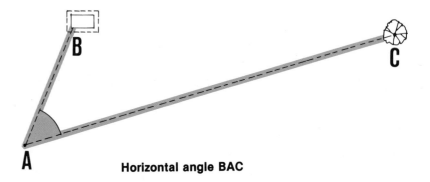

Horizontal angle BAC

3.   Each degree is divided into smaller units:

- 1 **degree** = 60 minutes (60′);
- 1 **minute** = **60 seconds (60″).**

These smaller units, however, can only be measured with high-precision instruments.

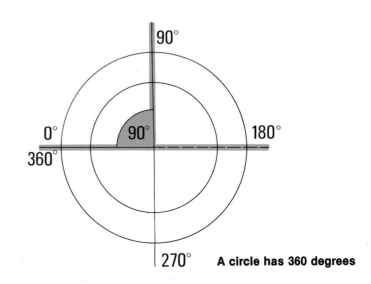

A circle has 360 degrees

Some general rules about angles

4.   A rectangular or a square shape has four straight sides and four interior 90° angles. The sum of these four interior angles is equal to 360°.

5.   The sum of the four interior angles of any four-sided shape is also equal to 360°, even if they are not right angles.

6.   It will be useful for you to remember the general rule that the **sum of the interior angles of any polygon** (a shape with several sides) **is equal to 180° times the number of sides, N, minus 2,** or:

Sum angles = (N − 2) × 180°

Examples

(a) A piece of land has five sides. The sum of its interior angles equals (5-2) x 180° = 540°.

(b) A piece of land has eight sides. The sum of its interior angles equals (8-2) x 180° = 1 080°.

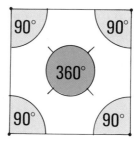 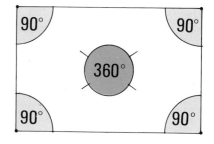

90° + 90° + 90° + 90° = 360°   90° + 90° + 90° + 90° = 360°
4 sides = 360°   **4 sides = 360°**

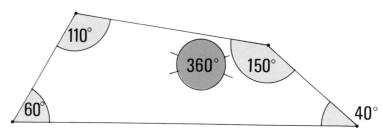

60° + 110° + 150° + 40° = 360°
4 sides = 360°

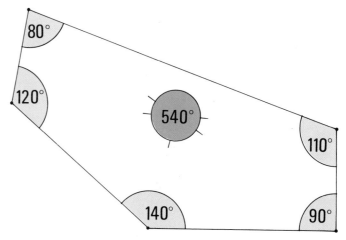

120° + 80° + 110° + 90° + 140° = 540°
5 sides = 540°

77

7. When you measure angles in a field, you can check on the accuracy of your measurements by applying this basic rule. Remember that **the sum of the interior angles of any triangle equals** $(3-2) \times 180° = 180°$.

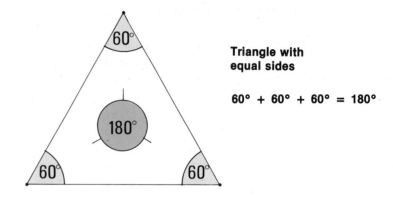

Triangle with equal sides

$60° + 60° + 60° = 180°$

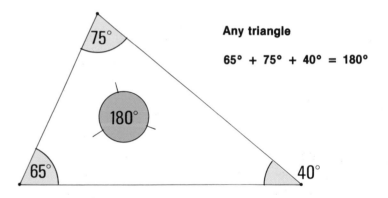

Any triangle

$65° + 75° + 40° = 180°$

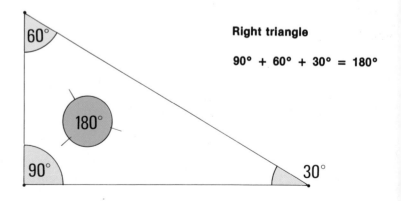

Right triangle

$90° + 60° + 30° = 180°$

Choosing the most suitable method

8.   There are only a few ways to **measure horizontal angles** in the field. The method you use will depend on how accurate a result you need, and on the equipment available. **Table 2** compares various methods and will help you to select the one best suited to your needs.

Note: because 90° angles are very important in topographical surveys, the measurement of these angles (used for laying out perpendicular lines) will be discussed in detail.

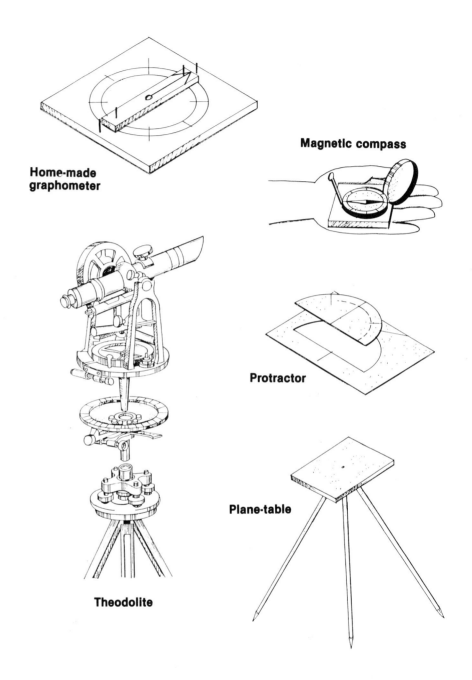

Home-made graphometer

Magnetic compass

Protractor

Theodolite

Plane-table

TABLE 2

Horizontal angle measurement methods

Section [1]	Method	Horizontal angle	Accuracy	Remarks	Equipment[2]
31 *	Home-made graphometer	Medium to large	Low	Best for 40-80 m For angles greater than 10°	*Graphometer*
32 **	Magnetic compass	Medium to large	Medium	Best for 40-100 m For angles greater than 10° No magnetic disturbance	Compass
33 *	Compass or protractor	All sizes	Low to medium	Dry weather only	Simple compass, *protractor*, drawing sheet
33 **	Plane-table	All sizes	Low to medium	Dry weather only	*Plane-table* drawing paper
34 *	Right-angle method	Small	Medium to high	Perpendicular to be set out	*Measuring line*
35 ***	Theodolite or transit	All sizes	High	Useful on long distances	Transit level with graduated horizontal circle
36	Miscellaneous	Right-angle only	Medium to high	Adapt method to length of perpendicular	Various

[1] * Simple   ** more difficult   *** most difficult.

[2] *In italics*, equipment you can make yourself from instructions in this book.

31   How to use the graphometer

1.   A graphometer is a topographical instrument used to measure horizontal angles. It is made up of a circle graduated in 360° degrees. Around the centre of this circle, a sighting device can turn freely. This device, called **an alidade**, makes it possible to create a line of sight that starts from your eyes, passes through the centre of the graduated circle, and ends at the selected landmark or ranging pole. When in use, the graphometer is rested horizontally on a stand.

2.   You can build your own graphometer by following the instructions below. It might be a good idea to ask a carpenter to help you.

Building your own graphometer

3.   Begin building your graphometer with the **graduated** circle given in **Figure 1**. You can make a photocopy of it, or you can draw a copy of it using tracing paper, or you can cut the page from the manual along the dotted line.

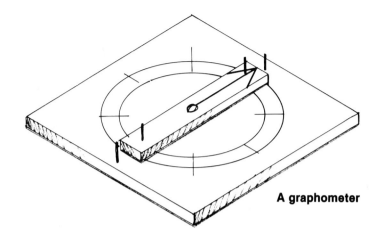

A graphometer

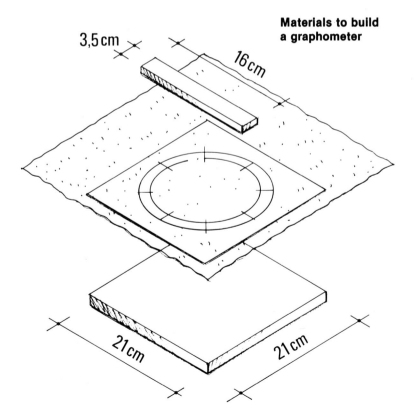

Materials to build a graphometer

3,5 cm

16 cm

21 cm

21 cm

81

4. Get a **wooden board**, 1 cm thick and 22 cm square.

5. Find its centre by drawing two diagonal lines on this board, from opposite corners. The point where the lines cross is the **exact centre** of the board.

6. Obtain one nut and bolt, about 1.5 cm long. At the **centre of the board, drill a hole into which the bolt will fit tightly**. On the bottom side of the board, drill the outside of the hole slightly larger to fit the nut.

7. Cut a hole of about the same size exactly at the **centre of the graduated circle** (Figure 1). Glue this sheet of paper on to the wooden board. Carefully align the central holes in the board and the circle, and make sure the four sides of the sheet are parallel to the sides of the board. You can do this easily by matching the two diagonal axes you have drawn on the board with the circle graduations 45º, 135º, 225º and 315º, respectively.

8. If possible, protect the sheet of paper. To do this, get a piece of **transparent plastic sheet** bigger than the board, and stretch it over the front side of the board. Attach it at the back with several thumbtacks.

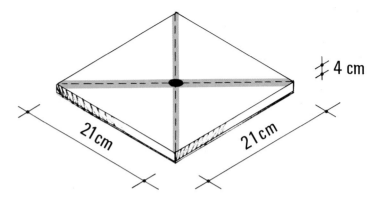

Find the centre of the board and drill a hole there

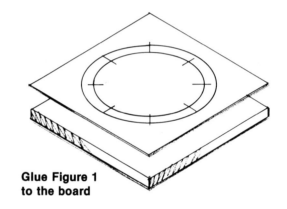

Glue Figure 1 to the board

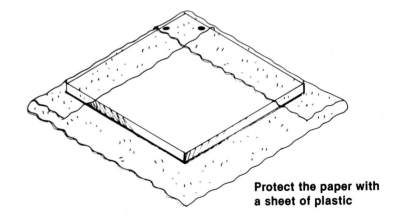

Protect the paper with a sheet of plastic

FIGURE 1

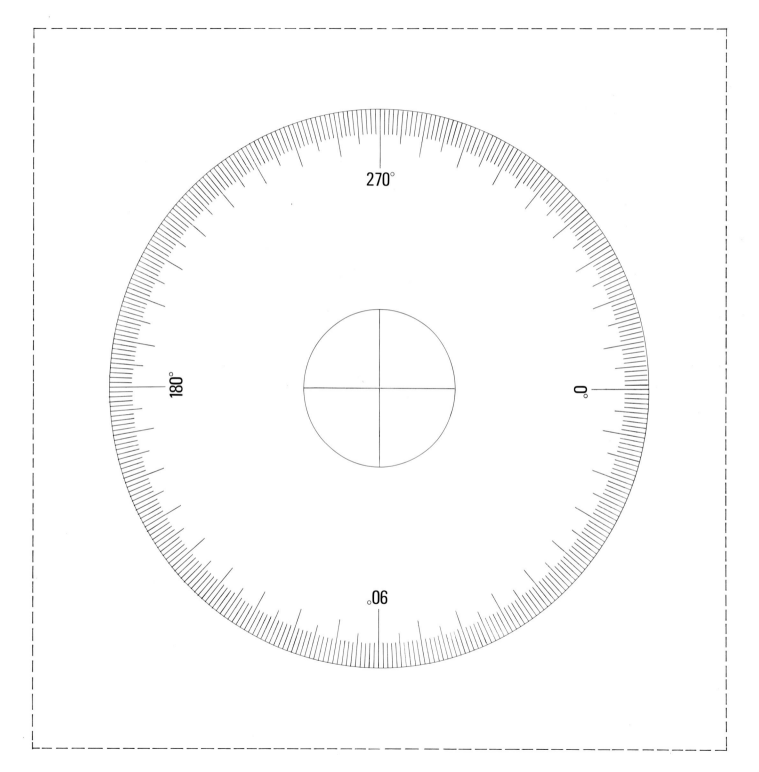

9. Now build the sighting device, called the **mobile alidade**. Get a thin wooden ruler, 16 cm long and 3.5 cm wide. Find its centre, as you did on the board, by drawing on it two diagonal lines from opposite corners. Draw a line through this centre point parallel to the long sides of the ruler. At the same point, drill a hole just a little larger than the diameter of the bolt. **Exactly** on the central line you have drawn, near each of its ends, drive a thin, **headless** nail 4 to 5 cm long into the ruler. Be careful not to let the nails break through to the other side of the ruler, and make sure you drive them in **vertically**. Your alidade is now ready.

10. To attach the alidade to the base, **place a thin washer** over the hole in the board you have prepared. Align the central hole of the alidade over the washer. Add a washer above the alidade and one below the wooden board, aligned with the central holes. Push the bolt through all the washers and holes and tighten the nut, so that the alidade will turn around under slight pressure.

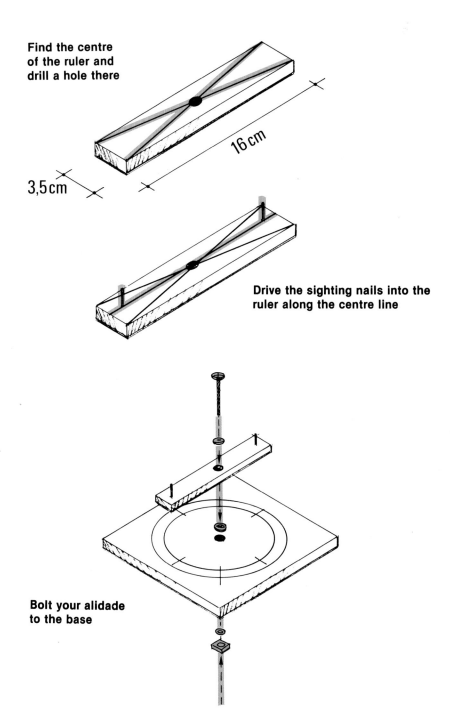

Find the centre of the ruler and drill a hole there

Drive the sighting nails into the ruler along the centre line

Bolt your alidade to the base

11.   On the board, along the 0° to 180° line, but outside the graduated circle, drive in two headless nails like the ones placed in the alidade. These will form a second line of sight. Clearly mark the top half of this line of sight with an arrow pointing exactly at the 0° graduation.

12.   On one end of the alidade, draw an arrow from the centre-bolt, along the median line, and through the nail at the end. The tip of the arrow should point exactly to the end of the median line above this nail. This arrow will help you to determine the graduation when you need to read it.

13.   To make your measurements more accurate, get a stake about 1.20 m high, and sharpen one end into a point. Drive this end into the ground and rest your graphometer on the other end as you measure.

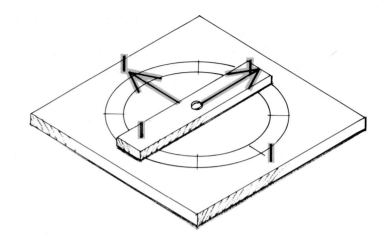

Mark your sighting line with nails and arrows

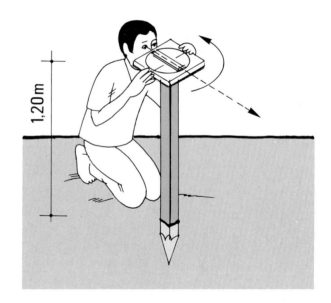

Using the graphometer with a stake to steady it

86

Using the home-made graphometer to measure horizontal angles

14.   Orient the graphometer with its **0° to 180° sighting line on the left side AB of the angle you need to measure**. Position the graphometer so that its **centre**, the bolt, is exactly above point A on the ground, the **station**, from which you will measure horizontal angle BAC. For more accuracy, you can use a **plumb-line** (see Section 48). If you have attached your graphometer by its centre to a stake, drive the pointed end of the stake **vertically** into the ground at the angle's summit A.

15.   Check that the graphometer is as horizontal as possible. To do this, place a round pencil on the board. If the pencil does not roll off, turn it 90° degrees and check again. When the pencil does not roll off in either direction, the graphometer is horizontal.

16.   Check again that **the 0° to 180° sighting line** is aligned well with the left side AB of the angle you need to measure. Adjust it if necessary, making sure not to disturb either the position of the station or the horizontality of the graphometer.

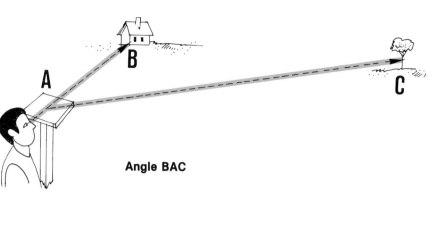

Angle BAC

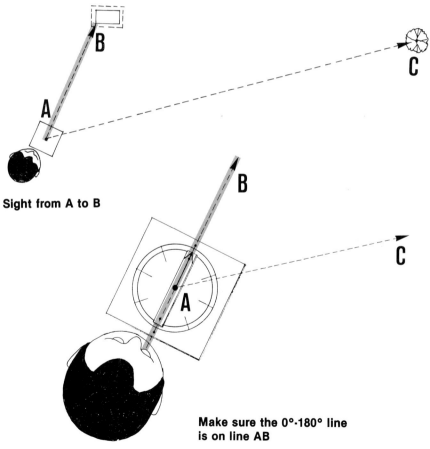

Sight from A to B

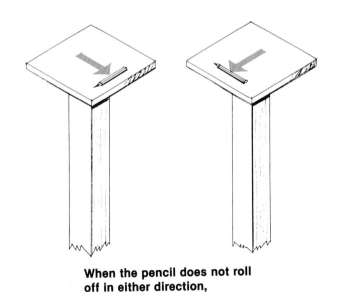

When the pencil does not roll off in either direction, the graphometer is horizontal

Make sure the 0°-180° line is on line AB

87

17. Move the **mobile alidade** around to the right until its sighting line lines up with the right side AC of the angle BAC.

18. Read the graduation above the arrow on the central line of the mobile alidade. This is the value of angle BAC in degrees.

Note: it is easier to position the graphometer above the station on the ground and make it horizontal without sighting with the 0° to 180° line. Just make sure that the **left side AB of the angle is to the right of this 0° to 180° line**. Then take two readings, using the mobile alidade for both the left side AB and the right side AC of the angle. The value of the angle equals the difference between these two readings.

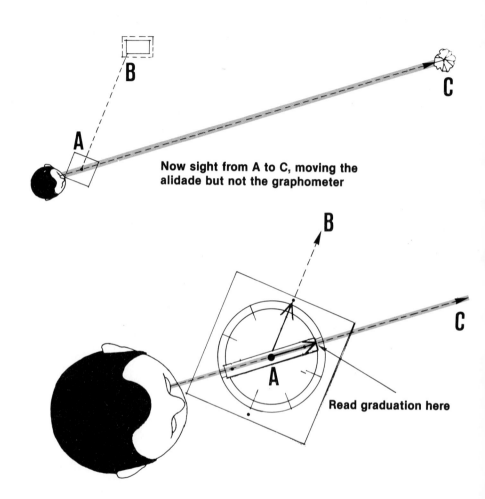

Now sight from A to C, moving the alidade but not the graphometer

Read graduation here

Example

Two lines, X and Y, form an angle XAY at station A.

- Clearly define lines X and Y by placing ranging poles at B and C, for example.
- Position the graphometer at station A, with its 0°-180° sighting line oriented to the left of AB.
- With the mobile alidade, sight at ranging pole B and read the graduation, AB = 23°.
- Turn the mobile alidade to sight at ranging pole C and read the graduation, AC = 75°.
- The angle BAC equals 75° − 23° = 52°.

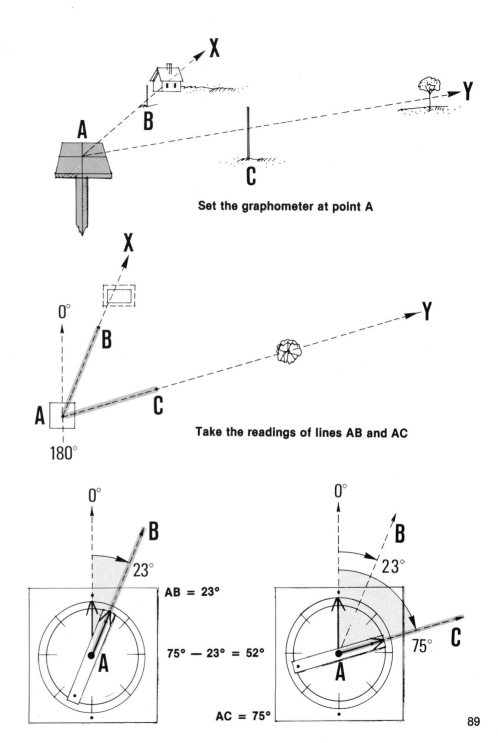

Set the graphometer at point A

Take the readings of lines AB and AC

AB = 23°

75° − 23° = 52°

AC = 75°

89

Measuring an angle with an inaccessible summit

19. To use the preceding method, you must be able to reach the summit A of the angle. When you cannot reach the summit, you have a choice of two methods.

20. You can **set out a line CB from any point** on one of the angle's sides to any point on its other side, making a triangle within the angle. Measure the two angles made by this new line and the angle's sides. The angle at the inaccessible summit of the triangle you have made equals the difference between 180° and the sum of the other two angles.

Example

You cannot reach summit A to measure angle XAY. From point B on line AX, set out line BC, where point C is on line AY. At station B, measure angle CBA = 60°; at station C, measure angle BCA = 73°. Calculate angle XAY = 180° - (60° + 73°) = 47°.

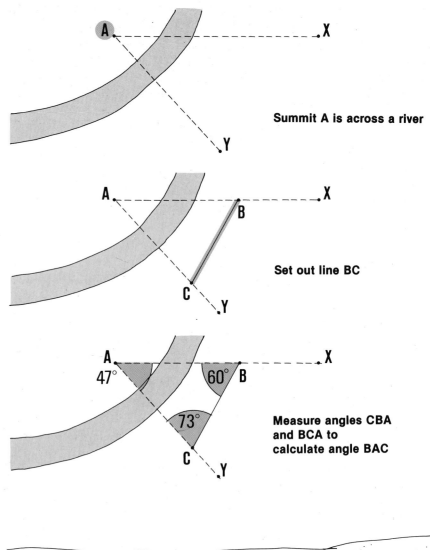

Summit A is across a river

Set out line BC

Measure angles CBA and BCA to calculate angle BAC

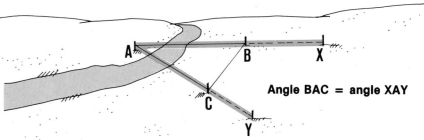

Angle BAC = angle XAY

21. Or you can lay out **two perpendicular lines** (see Section 36) from two points on one of the angle's sides. On each of these new lines, measure an equal distance. Join these two points with a line, which will be parallel to one of the angle's sides. Prolong the line until it intersects the other side of the original angle. At the intersection point, measure the new angle which is **equal to the summit angle.**

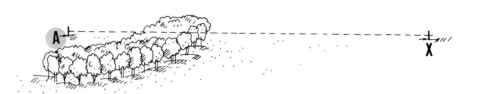

Summit A lies behind an obstacle

Example

You cannot reach summit A to measure angle XAY. On line AX, mark two points B and C. From these points, lay out perpendiculars BZ and CW. On these perpendiculars, measure segments of equal length from line AX, calling them segments BD and CE. Connect points E and D to form a line which is parallel to AX. Then extend line ED until it intersects line AY at point F. From a station at point F, measure angle EFY. Its measurement will be the same as that of angle XAY.

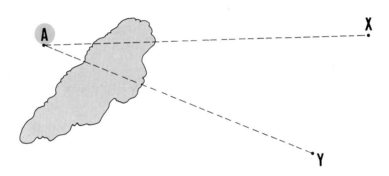

Set out perpendiculars BZ and CW

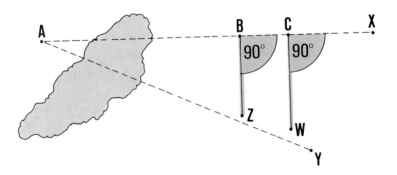

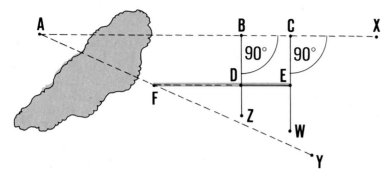

Find line ED, and prolong it to F

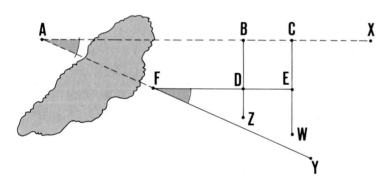

Measure angle EFY

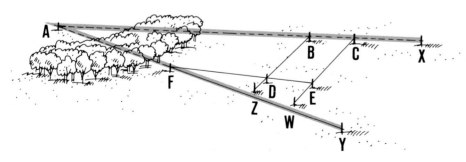

Angle EFY = angle XAY

Measuring consecutive angles

22. At one station, you may have to measure several angles formed by a series of lines which meet at one point, called **converging lines**. The angles they form are called **consecutive angles**.

23. To measure consecutive angles from one station, first measure all the angles by using the line furthest left as **the reference line**. Then, by simple subtraction, you can calculate the individual angles.

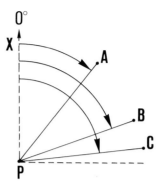

**XPA, APB and BPC
are consecutive angles**

Example

At station P, you have to measure three consecutive angles, XPA, APB and BPC. Take PX (furthest to the left) as the reference line and align the 0° graduation of the graphometer with it. Keeping the graphometer fixed in that position, move the mobile alidade around and measure each angle in turn (in this case, angles XPA = 40°, XPB = 70° and XPC = 85°). Calculate the consecutive angles as follows: XPA = 40°, as directly measured; APB = XPB − XPA = 70° − 40° = 30°; BPC = XPC − XPB = 85° − 70° = 15°.

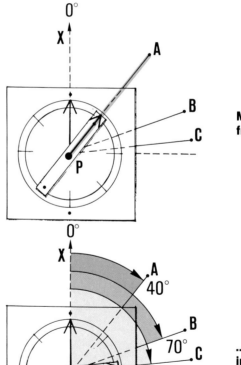

**Measure each angle
from the 0° line**

**...then calculate the
individual values**

32 How to use the magnetic compass

What is a magnetic compass?

1.   A simple magnetic compass is usually a **magnetic needle** which swings freely on a pivot at the centre of a **graduated ring**. The magnetic needle orients itself towards the **magnetic north***. The needle is enclosed in a case with a transparent cover to protect it.

2.   **Orientation compasses** are often mounted on a small rectangular piece of hard transparent plastic. They have a sighting line in the middle of a movable mirror. If you tilt this mirror, you can see both the compass and the ground line.

A simple compass

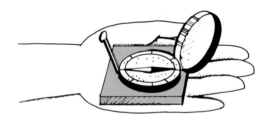

An orientation compass

3. **Prismatic compasses** give more accurate readings. To use one, hold it in front of your eyes so that you can read its scale. You can see the scale through a lens by means of **a prism**. Then turn the compass horizontally until the cross-hair is aligned with the ground mark (an optical illusion makes the hairline appear to continue above the instrument's frame). At the same time, the reading is shown on the compass-graduated circle behind the actual hairline. Since the graduated ring automatically orients itself, the reading directly gives **the measure of the angle between magnetic north* and the line of sight** (see also next paragraphs).

A prismatic compass

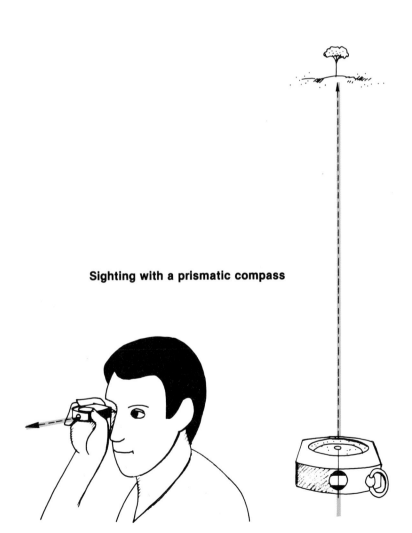

Sighting with a prismatic compass

4. A **magnetic needle** always points in the same direction - the **magnetic north**. This is why compasses are often used for orientation in the field and for mapping surveys (see, for example, Section 71 in Book 2 of this manual). The part of the compass needle pointing to the magnetic north is clearly marked, usually in red or a dark colour.

5. The **outside ring** of a compass is usually graduated in **360°**. **The 0° or 360° graduation** is marked **N**, which means **north**. In most compasses, the **graduation increases clockwise** and the following letters can be read on the circle:

- at 90°, **E** for **east**;
- at 180°, **S** for **south**;
- at 270°, **W** for **west**.

Intermediate orientations, such as NE, SE, SW and NW, are also shown sometimes.

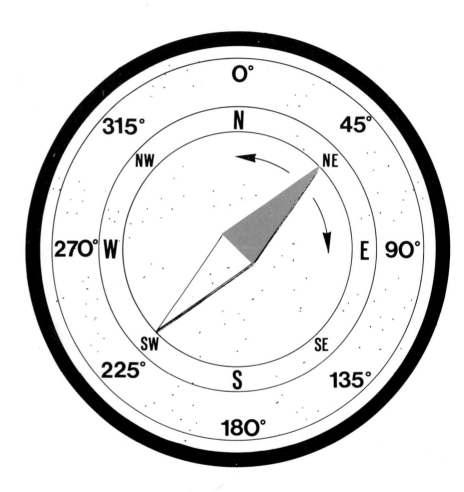

Using the compass to measure horizontal angles

6. You have learned that the needle of a compass always points in the same direction - the **magnetic north**. To use this direction as a reference, you need to make sure that the 0° graduation lines up with it. If the 0° graduation of your compass does not line up exactly with the magnetic north, turn the external ring until it does. Only then can you use your compass as described below.

7. At any station, the angle formed by the magnetic north and a straight line is called **the azimuth** of that line. This magnetic azimuth from north, called azimuth or Az, is always measured **clockwise, from the magnetic north to the line**.

Example

Azimuth OA = 37°; Az OB = 118°; Az OC = 230°; Az OD = 340°.

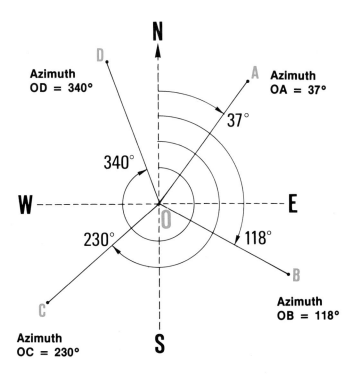

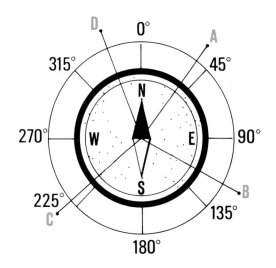

Measuring the azimuth of a line

8. To measure the azimuth of a line, take a position at any point on the line. Holding your compass horizontally, sight at another point on the line, such as a ranging pole. To do this, align the sighting marks of the compass with this point. If necessary (as with some orientation compasses), first align the 0° graduation N exactly with the northern point of the magnetic needle. At the intersection of the sighting line and the graduated ring, read the azimuth of the line from the point of observation. The reading will be most accurate if you **limit the length of the sighting line to 40 to 120 m**. Place more ranging poles on the line as you need them.

Note: to check the value of an azimuth, turn around and look in the opposite direction at another point on the same line. Read the measurement of this azimuth, which should **differ by 180°**. Usually, the difference will not be exactly 180°. If the difference is small enough, you can ignore it, or correct it by averaging the two readings. If it is large, you have made an error which you should correct (see later, **Main causes of error**).

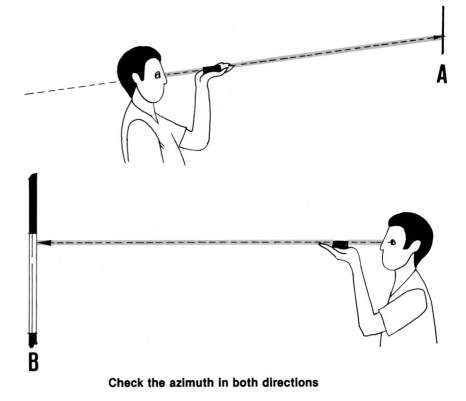

Check the azimuth in both directions

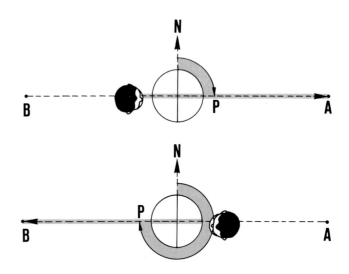

Example

To determine the azimuth of line XY, marked by ranging poles B and C, stand with the compass at station A in the middle of the line. Sight forward with the compass at ranging pole B and read azimuth AB = 65°. This is the **forward azimuth**. Check this value by turning around; sight backward with the compass at ranging pole C and read the **rear azimuth**, AC = 245°. The difference between the two azimuths is 245° – 65° = 180°, which means that the measurements are accurate.

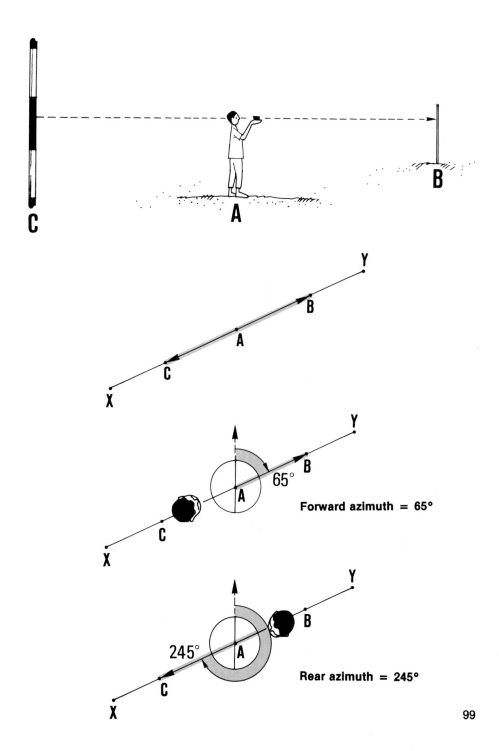

Forward azimuth = 65°

Rear azimuth = 245°

Measuring a horizontal angle

9.   To measure a horizontal angle, stand at the angle's summit and measure the azimuth of each of its sides; calculate the value of the angle as follows.

10.   If the magnetic north falls outside the angle, calculate the value of the angle between the two lines of sight as equal to **the difference between their azimuths**. Always subtract the smaller number from the larger one, no matter which azimuth you read first. Just be sure magnetic north is not inside the angle.

Example

(a) Angle BAC; Az AB = 25⁰; Az AC = 64⁰; BAC = 64⁰ – 25⁰ = 39⁰

(b) You have to measure angle XAY; measure the azimuth of AX = 265⁰; measure the azimuth of AY = 302⁰; the angle XAY measures 302⁰ – 265⁰ = 37⁰.

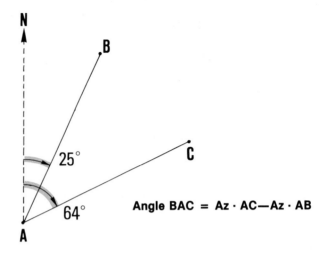

Angle BAC = Az · AC—Az · AB

Angle XAY = Az · YA—Az · XA

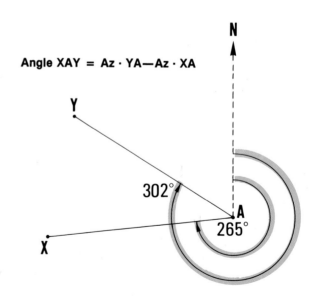

11.   If the magnetic north falls inside the angle, the angle between the two lines of sight is equal to **360° minus the difference between their azimuths**. To calculate the angle, first find the difference as you did in step 10, above, then subtract this number from 360°.

Example

You have to measure angle EAF; measure the azimuth of AE = 23°; measure the azimuth of AF = 310°; angle EAF measures 360° – (310° – 23°) = 73°.

Note: to check on your measurements and to improve their accuracy, you should repeat each measurement three times from the same station. These measurements should give similar results.

12.   If you cannot reach the summit of the angle, separately measure the azimuth of each line from another point situated on it (see step 8, above) and calculate the angle as in step 9, above.

Example

You have to measure angle BAC, but the summit A is not accessible; at point X on AB, measure azimuth XB = 39°; at point Y on AC, measure azimuth YC = 142°. Since the magnetic north falls outside angle BAC, calculate its measurement 142° – 39° = 103°.

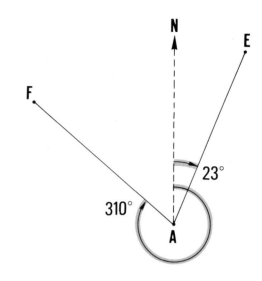

Angle EAF = 360° — (Az · AE — Az · AF)

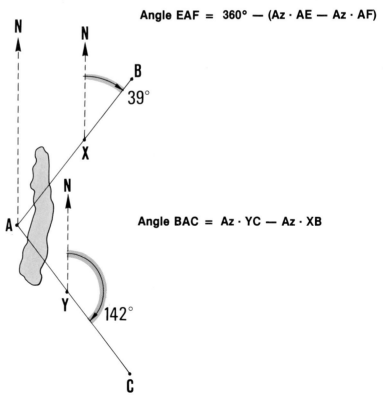

Angle BAC = Az · YC — Az · XB

13.   When you must survey a polygonal* site, measure the azimuth of two sides from each of the summits. For each side of **the polygon**, you will thus determine **one forward and one rear azimuth**. You can then check on the accuracy of the two azimuths, which should differ by 180º. If they do not, subtract 180º from the greatest azimuth and calculate the average between this value and the smallest azimuth. To do this, add the two numbers and divide by two. From averages like this for the other pairs of azimuths, you can calculate the interior angles of **the polygon**, as explained above.

Note: to make a final check, add all the interior angles. This sum should equal (N − 2) × 180º N being the number of sides of **the polygon** (see Section 30, step 6).

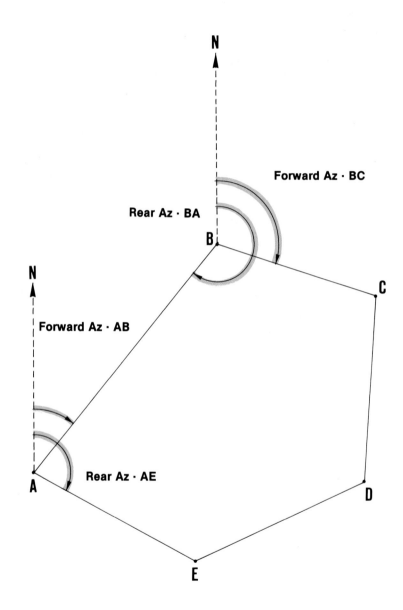

Example

You have to survey polygon ABCDEA. From summit A measure forward Az AB = 40° and rear Az AE = 120°. Move clockwise to summit B and measure rear Az BA = 222° and forward Az BC = 110°. Proceed in the same way from the other three summits C, D and E. In total, you get ten measurements. Mark them down in your notebook. (See **columns** 1 and 2 where the order of measurements is shown in parentheses.)

Calculate the values of **column** (3) by subtracting 180° from the largest azimuth measured at each summit. This gives you values which should be **equal to the smallest observed azimuths**, written either in column 1 or in column 2, according to the position of the summit.

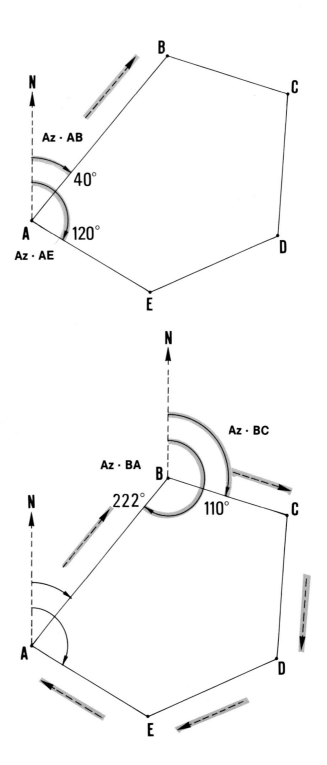

- When the values are equal to the smaller observed azimuths (summits C, E), transfer these measurements to columns 4 or 5, according to the type of azimuth they represent.
- When they are not equal (summits A, B, D):

  - Use columns 1 or 2 and column 3 to calculate the average **smallest azimuth**. To do this, add the measurement of the smallest Az from column 1 or 2 to the number in column 3. Divide the total by 2 to find the average. For example at summit A, forward Az AB = (42 + 40) ÷ 2 = 41°. At summit D, rear Az ED = (66 + 68) ÷ 2 = 67°. Enter a **forward** Az in column 4 and a **rear** Az in column 5.
  - Add 180° to the smallest **calculated azimuths** to calculate the remaining azimuths. For example, at summit A, rear Az BA = 41 + 180 = 221° and at summit D, forward Az DE = 67 + 180 = 247°. As before, enter a forward Az in column (4) and a rear Az in column (5).

Summit of Polygon	Observed Azimuths		Largest Azimuth − 180°	Calculated Azimuths	
	Forward Az	Rear Az		Forward Az	Rear Az
Column	1	2	3	4	5
A	(1)A B = 40	(4) BA = 222	42	AB = 41	BA = 221
B	(3) BC = 110	(6) CB = 288	108	BC = 109	CB = 289
C	(5) CD = 185	(8) DC = 5	5	CD = 185	DC = 5
D	(7) DE = 246	(10) ED = 68	66	DE = 247	ED = 67
E	(9) EA = 300	(2) AE = 120	120	EA = 300	AE = 120

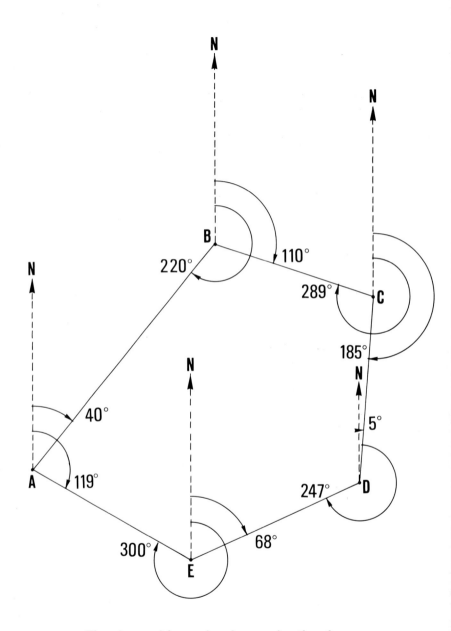

The observed forward and rear azimuths of polygon ABCDE, in a table and in a sketch

- Calculate **interior angles**, combining the calculated azimuths (columns 4 and 5) two by two as follows, with the help of a little sketch:
  - angle EAB = Az AE – Az AB = 120° – 41° = 79° .angle ABC = Az BA – Az BC = 221° – 109° = 112°
  - angle BCD = Az CB – Az CD = 289° – 185° = 104°
  - angle CDE = 360° – (Az DE – Az DC) = 360° – (247° – 5°) = 118°
  - angle DEA = 360° – (Az EA – Az ED) = 360° – (300° – 67°) = 127°.

- Check your calculations: the sum of the angles should be equal to (5 – 2) × 180° = 540°. These calculations (79° + 112° + 104° + 118° + 127° = 540°) are correct.

14. If you have to measure consecutive angles, proceed as described earlier (see end of Section 31).

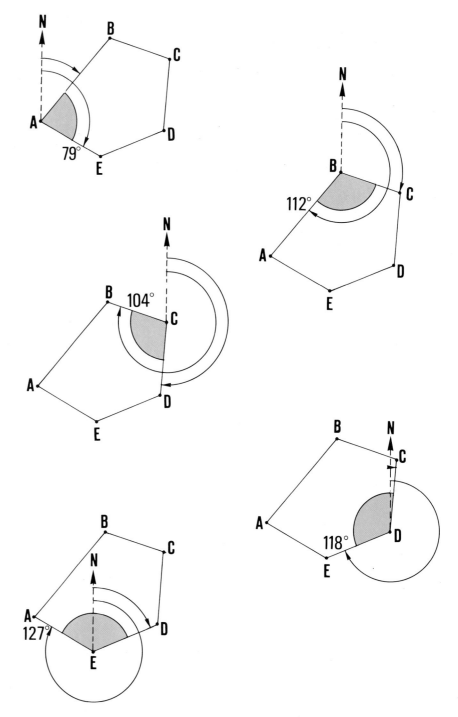

Checking when using a compass

When using a magnetic compass to measure horizontal angles, you should carefully check the following points:

15. ● The magnetic needle must swing freely on its pivot. Keep the compass horizontal in one hand and, with the other hand, bring an iron object close to the magnetized needle's point. Make the needle move to the left with the iron; when you move the iron away, the needle should swing quickly and smoothly to its original position. **Repeat** the movement in the opposite direction to double check.

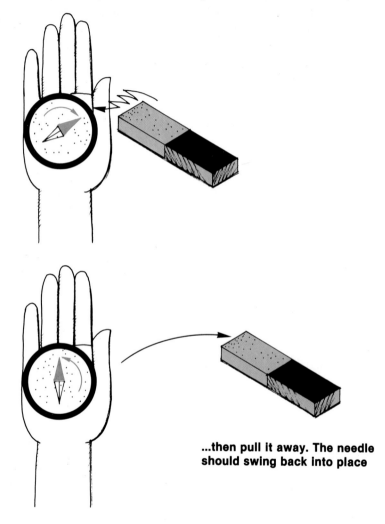

Put the magnet near the compass to attract the needle

...then pull it away. The needle should swing back into place

16. ● The magnetic needle must be horizontal when the compass is horizontal. Lay the compass on a horizontal wooden surface (such as a table) and check that the needle remains horizontal. If it does not, you will have to open the case of the compass and add a light weight to the needle. To do this, you can wind some cotton sewing thread around the part of the needle that is highest, and move the needle back and forth until it is balanced and horizontal.

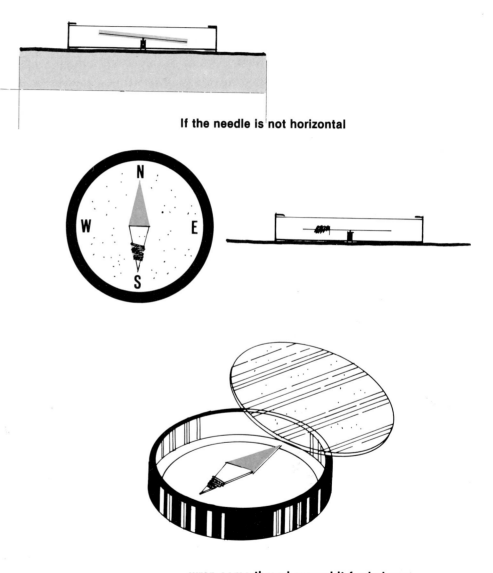

If the needle is not horizontal

...wrap some thread around it for balance

17. ● Do not keep iron objects close to the compass. Iron will attract the magnetic needle, and your measurements will be wrong. Distance **measuring lines** made of metal, such as steel bands, steel tapes and chains, as well as metal ranging poles and marking pins, should be kept 4 to 5 m away from the compass when you are measuring angles. If you wear **eyeglasses with metal frames**, you will also have to keep them away from the compass. Remember that concrete structures (towers, bridges, etc.) are built with iron bars which may also cause the compass needle to move.

18. ● Do not use a compass when there is **thunder**. It affects the needle.

19. ● Do not use a compass near an eletric **power line**.

20. ● Keep the compass horizontal while you are measuring with it.

Note: because the magnetic needle of the compass is **always** affected by the presence of iron nearby, **checking the measured azimuths** (as explained earlier) is extremely important. If your results do not agree after repeated measurements, local **magnetic disturbances** caused by the presence of iron in the ground may be responsible for the errors. You should then use another method of measurement.

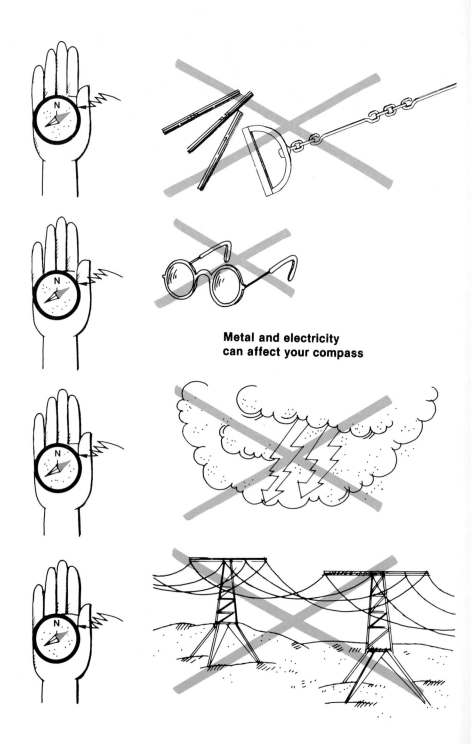

Metal and electricity can affect your compass

108

33 Graphic methods for measuring horizontal angles

To use the graphic methods for measuring horizontal angles, you need to draw the angle on paper first. Then you will measure the angle with a **protractor** (see step 11, below). As you have seen with other methods, you can obtain more accurate results if you repeat the procedure at least twice to discover possible errors.

Using a simple compass and a protractor in the field

1. With this method, you can use a **simple magnetic compass** (see Section 32.) The only purpose of this compass is to show the direction of the **magnetic north***.

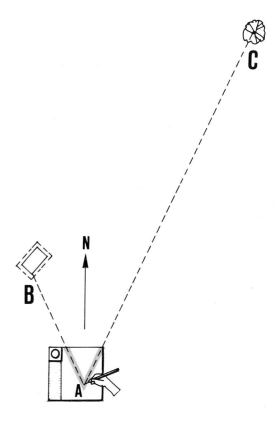

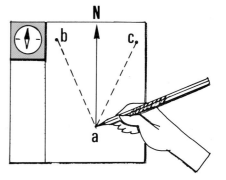

2. Get a 30 × 30 cm piece of stiff cardboard or thin wooden board, and several sheets of **square-ruled paper** (such as millimetric paper). Lightly glue each sheet, at its four corners, to the board, one on top of the other.

3. On the upper left-hand corner of the top sheet, attach the **compass**, for example with a string or rubber band or within a small wooden frame, so that its **0° to 180° reference line** is parallel to one of the rules on the paper. With a pencil, draw an arrow straight up toward the top of the sheet, and mark it North.

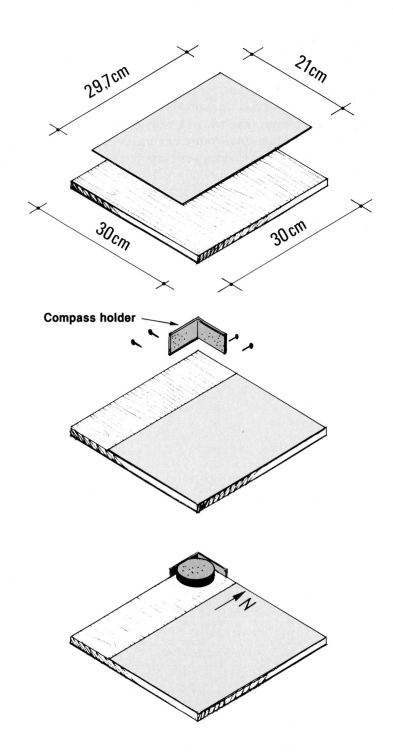

4. To draw the horizontal angle BAC you need to measure, stand at the **angle's summit A** and look at a ground line AB which forms one of the sides of the angle.

5. Keeping your board horizontal on the palm of one hand held in front of you, turn it slowly around until the northern point of **the compass needle reaches the 0°-graduation**. Your sheet of paper is now oriented, with its arrow facing north.

Note: it will be easier if you rest the board on a stable support, such as a wooden pole driven into the ground.

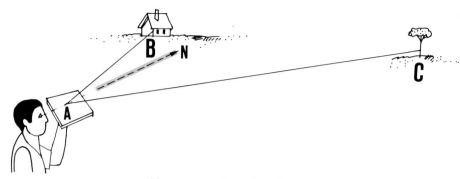

Line up your board so the compass points to magnetic north

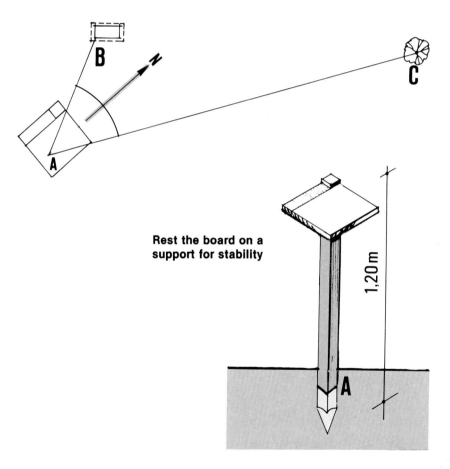

Rest the board on a support for stability

1,20 m

111

6. Without moving the board any more, trace on the paper in pencil, with your free hand, a line ab going straight ahead in the direction of the ground line AB.

7. Repeat the procedure described in steps 5 and 6 above, looking at the ground line AC which forms the other side of the angle and drawing line ac.

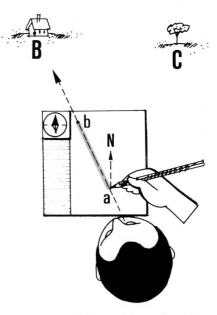

Sight and draw line ab

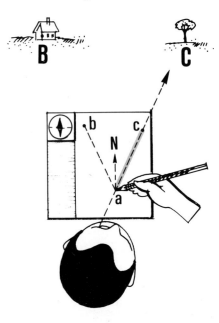

Then sight and draw line ac

8. Using a protractor (see steps 15 to 17 below), measure the azimuths of the lines you have traced as the **angles formed by them with any of the paper rules running parallel to the north**. Remember to measure the angle clockwise from the north to the pencil line (see Section 32).

Note: you only need to measure angles smaller than 90°, since the square-ruled paper shows the 90°, 180° and 270° directions.

9. Take the azimuths of the two sides of the horizontal angle, and calculate the value of the angle as described in Section 32.

Using a plane-table and a protractor

10. If you have a **plane-table** (see Section 75), you can use it to draw the angles on paper while you are in the field. Then it is easy to measure them with a protractor (see steps 15 to 17, below).

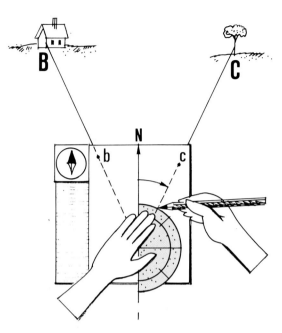

Measure the azimuths with a protractor

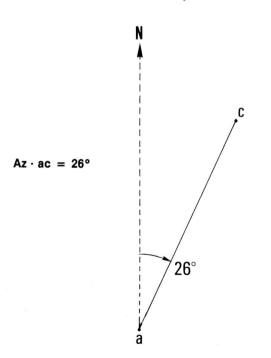

Az · ac = 26°

11. A **protractor** is a small instrument used in drawing. It is graduated in **degrees** or **fractions of degrees**. **The semicircular protractor** is the most common type, but a **full-circle protractor** may be best for measuring angles greater than 180°. Protractors are usually made of plastic or even paper. You can buy one cheaply in stores that sell school supplies. Or you can use the one provided in **Figure 2**. Either make a photocopy of it, or copy it on to transparent tracing paper, or cut it from the manual. Note that the **arrow** at point A indicates the exact position of the **protractor's centre**.

FIGURE 2

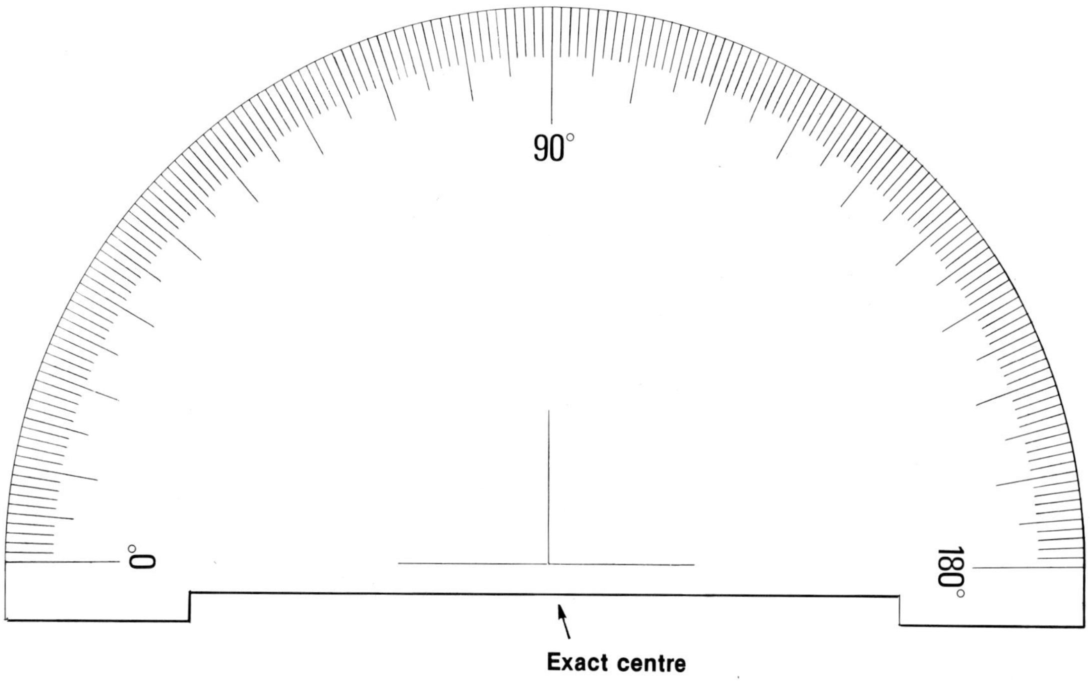

90°

0°

180°

Exact centre

115

Making your own protractor

12.   Cut the drawing of the protractor in Figure 2, or your copy of it, exactly along its curved outer edge.

13.   Glue this paper protractor onto a slightly larger piece of stiff cardboard.

14.   Cut the cardboard, exactly following the shape of the protractor.

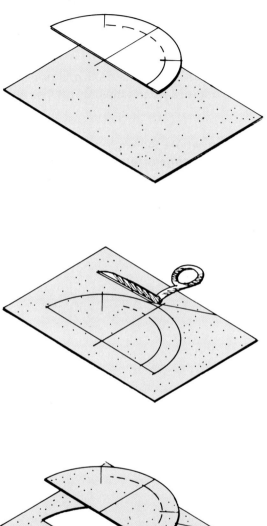

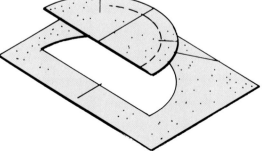

Using the protractor to measure an angle you have drawn

15.   Align the **0° to 180° straight edge** of the protractor with one of the angle's sides AB.

16.   Move the protractor so that its centre is positioned **exactly** on the summit A of the angle, keeping the 0° to 180° straight edge on the angle's side AB.

17.   Look for the point where the angle's other side AC intersects the graduation on the **round edge** of the protractor. Read the value of the angle from this graduation. This value may be expressed in **degrees** and **minutes** (remember that half of 1 degree equals 30 minutes).

Note: if the sides of the angle are not long enough to intersect the edge of the protractor, lengthen them before you begin measuring.

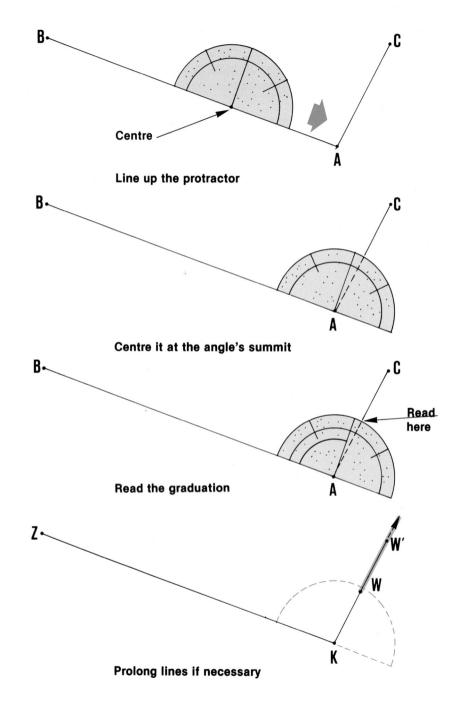

Centre

Line up the protractor

Centre it at the angle's summit

Read here

Read the graduation

Prolong lines if necessary

34 How to measure horizontal angles by the right-angle method

1.   The right-angle method is best for measuring **angles smaller than 10 degrees** in the field, since the preceding methods do not give accurate results. The right-angle method is based on the geometrical properties of right-angled triangles (see Section 30, step 7).

2.   From the angle's summit A, **measure 10 m** along one of the sides AC of the angle. Clearly mark this point D, with a ranging pole for example.

3.   From point D, lay out a **perpendicular line** and prolong it until it intersects the second side of the angle. Clearly mark this intersection point E.

4.   Accurately measure the length, in metres, of this perpendicular line DE.

5.   Divide this length by 10 to obtain the **tangent*** of the angle.

6.   Look for this value in **Table 3** and find the measurement of the angle BAC in degrees and minutes.

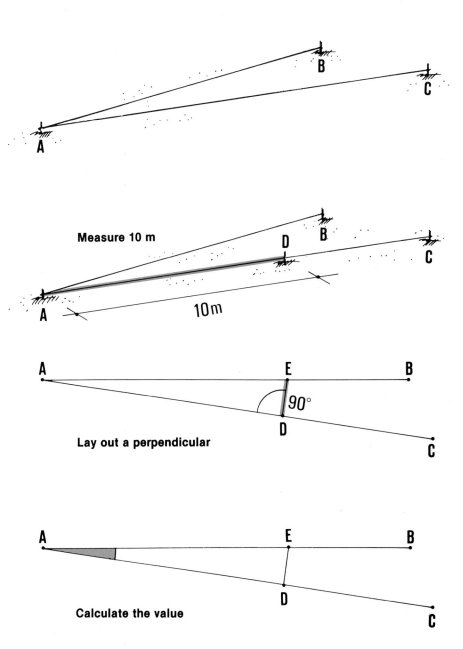

Measure 10 m

10m

Lay out a perpendicular

Calculate the value

Example

You have to measure the small angle XAY:

- from summit A measure 10 m on line XA and mark point B;
- from B trace perpendicular line BZ which intersects line YA at point C;
- exactly measure distance BC = 1.12 m;
- dividing this value by 10, you obtain the tangent of angle XAY, 0.112;
- look for 0.112 in Table 3. The closest value you find is 0.1110. Based on this value, angle XAY = 6°20'.

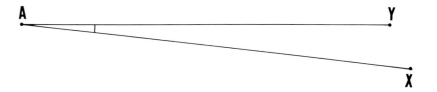

Measure 10 m to point B

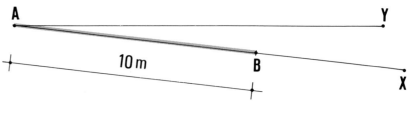

Lay out perpendicular BZ

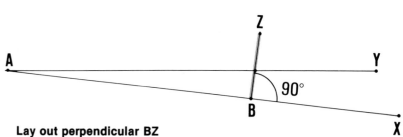

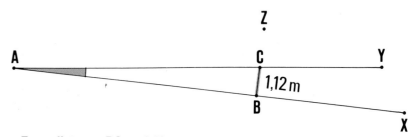

From distance BC = 1.12 m, calculate the value XAY = 6°20'

TABLE 3

Tangents and values of angles
(*Tan* = angles expressed in degrees *d* and minutes *m*)

Tan	d	m	Tan	d	m	Tan	d	m	Tan	d	m	Tan	d	m	Tan	d	m	Tan	d	m
0	0	0	0.0875	5	0	0.1763	10	0	0.2679	15	0	0.3640	20	0	0.4663	25	0	0.5774	30	0
0.0029		10	0.0904		10	0.1793		10	0.2711		10	0.3673		10	0.4699		10	0.5812		10
0.0058		20	0.0934		20	0.1823		20	0.2742		20	0.3706		20	0.4734		20	0.5851		20
0.0087		30	0.0963		30	0.1853		30	0.2773		30	0.3739		30	0.4770		30	0.5890		30
0.0116		40	0.0992		40	0.1883		40	0.2805		40	0.3772		40	0.4806		40	0.5930		40
0.0145		50	0.1022		50	0.1914		50	0.2836		50	0.3805		50	0.4841		50	0.5969		50
0.0175	1	0	0.1051	6	0	0.1944	11	0	0.2867	16	0	0.3839	21	0	0.4877	26	0	0.6009	31	0
0.0204		10	0.1080		10	0.1974		10	0.2899		10	0.3872		10	0.4913		10	0.6048		10
0.0233		20	0.1110		20	0.2004		20	0.2931		20	0.3906		20	0.4950		20	0.6088		20
0.0262		30	0.1139		30	0.2035		30	0.2962		30	0.3939		30	0.4986		30	0.6128		30
0.0291		40	0.1169		40	0.2065		40	0.2994		40	0.3973		40	0.5022		40	0.6168		40
0.0320		50	0.1198		50	0.2095		50	0.3026		50	0.4006		50	0.5059		50	0.6208		50
0.0349	2	0	0.1228	7	0	0.2126	12	0	0.3057	17	0	0.4040	22	0	0.5095	27	0	0.6249	32	0
0.0378		10	0.1257		10	0.2156		10	0.3089		10	0.4074		10	0.5132		10	0.6289		10
0.0407		20	0.1287		20	0.2186		20	0.3121		20	0.4108		20	0.5169		20	0.6330		20
0.0437		30	0.1317		30	0.2217		30	0.3153		30	0.4142		30	0.5206		30	0.6371		30
0.0466		40	0.1346		40	0.2247		40	0.3185		40	0.4176		40	0.5234		40	0.6412		40
0.0495		50	0.1376		50	0.2278		50	0.3217		50	0.4210		50	0.5280		50	0.6453		50
0.0524	3	0	0.1405	8	0	0.2309	13	0	0.3249	18	0	0.4245	23	0	0.5317	28	0	0.6494	33	0
0.0553		10	0.1435		10	0.2339		10	0.3281		10	0.4279		10	0.5354		10	0.6536		10
0.0582		20	0.1465		20	0.2370		20	0.3314		20	0.4314		20	0.5392		20	0.6577		20
0.0612		30	0.1495		30	0.2401		30	0.3346		30	0.4348		30	0.5430		30	0.6619		30
0.0641		40	0.1524		40	0.2432		40	0.3378		40	0.4383		40	0.5467		40	0.6661		40
0.0670		50	0.1554		50	0.2462		50	0.3411		50	0.4417		50	0.5505		50	0.6703		50
0.0699	4	0	0.1584	9	0	0.2493	14	0	0.3443	19	0	0.4452	24	0	0.5543	29	0	0.6745	34	0
0.0729		10	0.1614		10	0.2524		10	0.3476		10	0.4487		10	0.5581		10	0.6787		10
0.0758		20	0.1644		20	0.2555		20	0.3508		20	0.4522		20	0.5619		20	0.6830		20
0.0787		30	0.1673		30	0.2586		30	0.3541		30	0.4557		30	0.5658		30	0.6873		30
0.0816		40	0.1703		40	0.2617		40	0.3574		40	0.4592		40	0.5696		40	0.6916		40
0.0846		50	0.1733		50	0.2648		50	0.3607		50	0.4628		50	0.5735		50	0.6959		50

35 How to measure horizontal angles with a theodolite

What is a theodolite?

1.   A **theodolite**, sometimes called a **transit**, is an expensive instrument which engineers use to measure horizontal angles precisely. It is like a graphometer, but more complicated (see Section 31). Most theodolites are designed to measure vertical angles as well (see Sections 47 and 59). The theodolite's basic features for measuring horizontal angles are:

- a **horizontal circle**, graduated in degrees, which may be rotated and then clamped in any position;
- a **circular plate** which may be rotated inside this circle, and which shows additional graduations for reading the graduations on the circle with greater precision;
- a **telescope** which is attached to this circular plate and turns with it, and which can also be turned up and down in the vertical plane.
- a **tripod** (three-legged support) on which to place the theodolite when measuring.

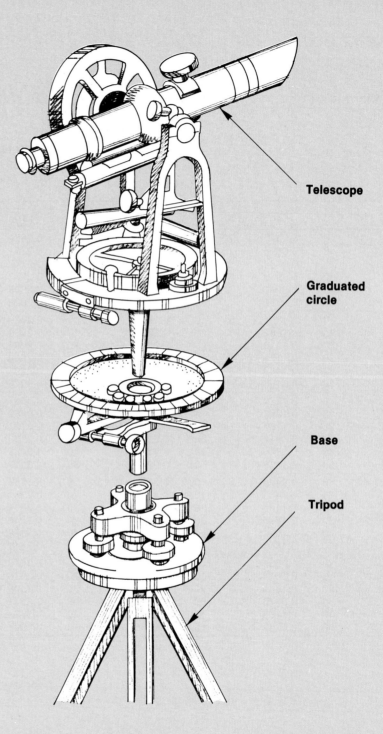

Telescope

Graduated circle

Base

Tripod

Using the theodolite to measure a horizontal angle

2. If you want to measure angle BAC, place the theodolite on its tripod at summit A. Set the index on the horizontal graduated circle at **zero** and take a sighting to B. Clamp the circle in position. Turn the telescope and its circular plate to take a sighting to C, while rotating through the angle BAC. You may read the angle measurement directly from the circular plate index.

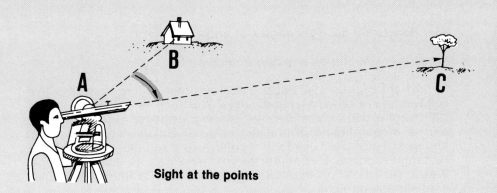

Sight at the points

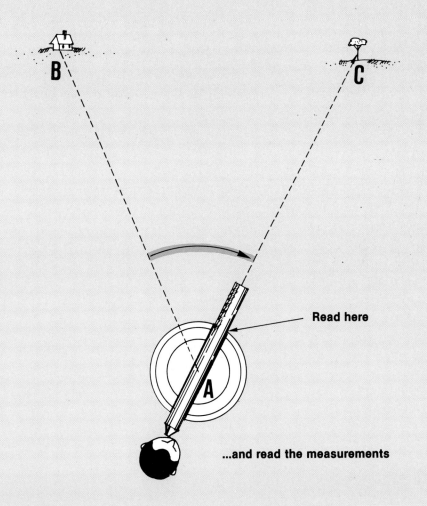

Read here

...and read the measurements

123

36 How to set out right angles or perpendiculars

What are right angles and perpendiculars?

1. A right angle is a **90° angle**. Two lines intersecting each other at a right angle are called **perpendiculars**. You have already learned how right angles can be useful for measuring distances (see Section 29) and for measuring horizontal angles (see Section 31, step 21). Right angles are also often used in fish culture, for example when you build rectangular ponds, to estimate the volume of a future reservoir (Vol.4, **Water**, Section 42), or measure land areas (see Chapter 10 in Book 2 of this manual).

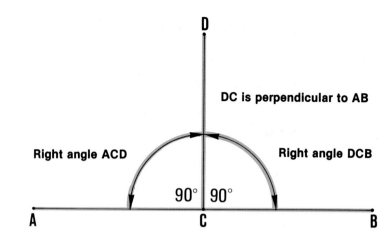

DC is perpendicular to AB

Right angle ACD   **Right angle DCB**

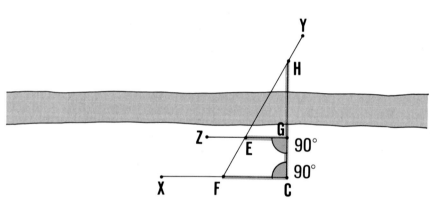

Some uses of perpendiculars

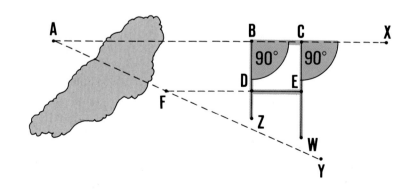

What are the main problems you will face?

2. There are two main problems to be solved:

    ● you have **to drop a perpendicular** from a given point A to a line XY; or
    ● you have **to lay out a perpendicular XY** to another line AB from a given point X.

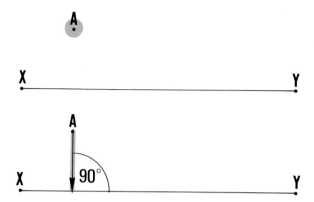

Dropping a perpendicular from point A to line XY

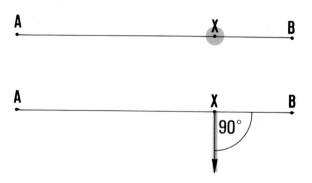

Laying out a perpendicular from point X on line AB

125

Dropping a perpendicular by the full-circle method

3. Set out line XY with ranging poles, and point A above or below it with a marking pin. You will drop the perpendicular from point A to XY. Get a simple line (a liana or rope) or a measuring line (a tape or chain) slightly longer than the distance from point A to line XY.

4. Attach one end of the line to the marking pin **at point A**, keeping it near the ground.

5. Walk with the other end of your measuring line to the line XY, and stop about **2 m beyond the point where you crossed it**.

6. Holding the line in your hand, **trace an arc** with it on the ground. To do this, swing the line in a curve to the left until you intersect XY, and mark this point B. Then swing the line in a curve to the right until you intersect XY, and mark this point C.

Put a marking pin at point A

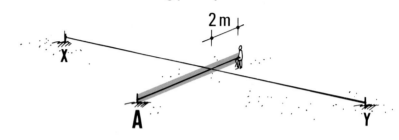

Attach the line to the pin

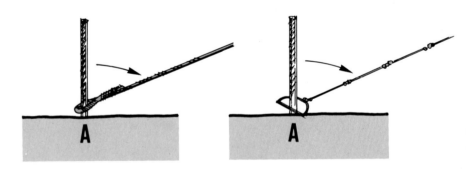

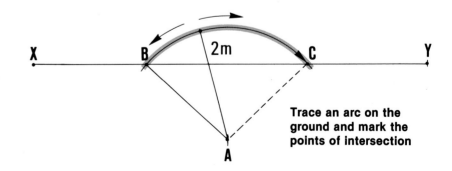

Trace an arc on the ground and mark the points of intersection

7. Measure on XY the **distance BC** between these two marked points.

8. Divide this distance by 2 and measure this new distance from point B. Mark this point D – it will be exactly in the middle of BC.

9. Connect point D and the original point A to form a new line **AD perpendicular to XY**.

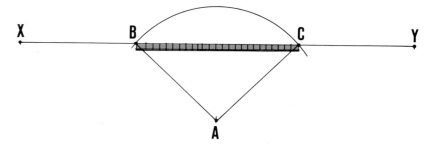

Measure distance BC

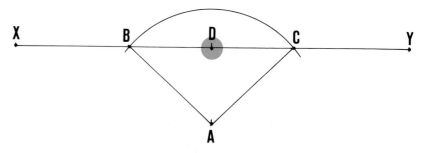

Divide by 2 to find mid-point D

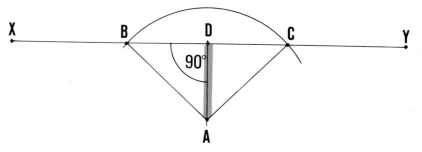

Connect points D and A to form the perpendicular

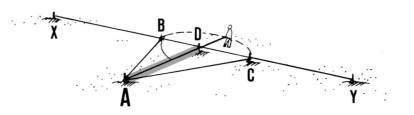

Dropping a perpendicular by the half-circle method

Set out line XY and point A as described above. Prepare a measuring line a little longer than half the distance from point A to the line XY on which you will drop the perpendicular.

10. From any point B on line XY, measure distance AB to point A.

11. Divide this distance AB by 2 and mark the centre point C.

12. Attach one end of your measuring line to point C as in step 4, above.

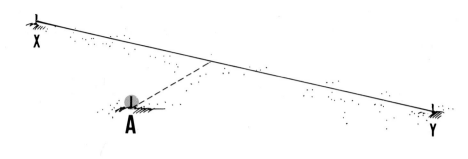

Mark point A

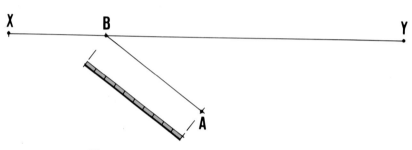

Measure distance AB

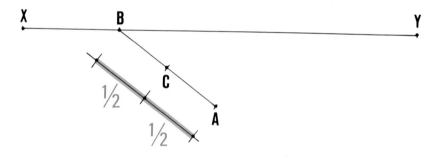

Divide by 2 to find mid-point C

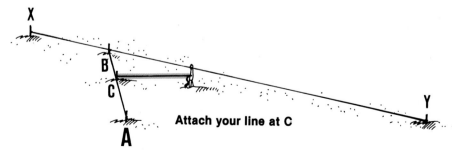

Attach your line at C

128

13. Walk with the other end of your line to point B on XY, and **clearly mark this distance** CB on the measuring line.

14. Trace an arc on the ground with the line length CB. To do this, swing the line in a curve to the right until you intersect XY again. Mark this point D.

15. Join D to the original point A to form a new line AD perpendicular to XY.

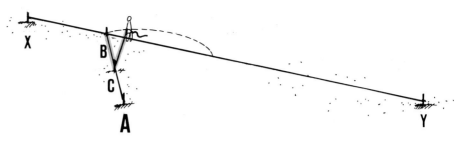

Find length CB on your rope

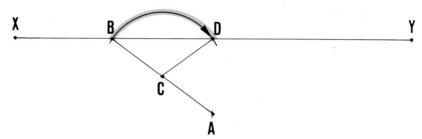

Trace an arc on the ground to find point D

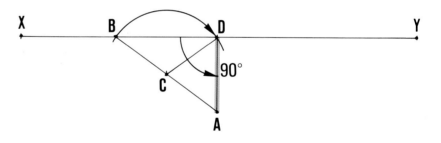

Connect point A and D to form the perpendicular

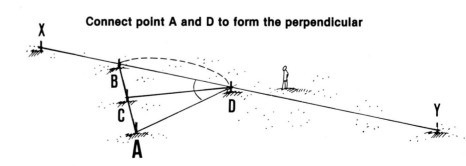

129

16.   The easiest way of setting out a perpendicular from a fixed point A on line XY is to use a **simple line clearly marked at its mid-point** with, for example, a knot. This line can be a liana, a rope or a string or you may use a measuring tape, whose graduations will help you to locate the exact mid-point. For good results, **the line should be at least 8 m long**. A longer line will make your measurements even more accurate. If you are working alone, make a small loop at each end of the line.

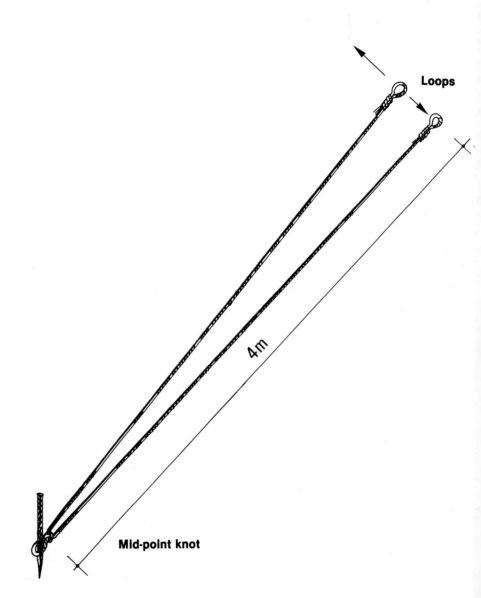

Loops

4m

Mid-point knot

17. Set out line XY and mark point A from which you will set out the perpendicular. On either side of point A and along line XY, measure the **equal distances AB** = **AC** of about 2 m. You can use part of the measuring line to do this. Mark points B and C with stakes.

18. Loop **one end of the line** over stake B and the other end over stake C.

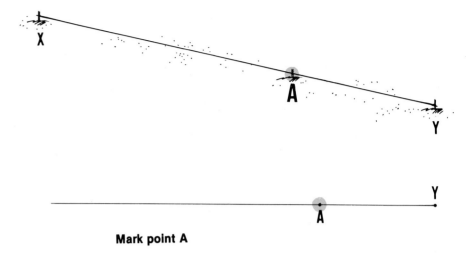

Mark point A

Measure 2 m each way to find points B and C

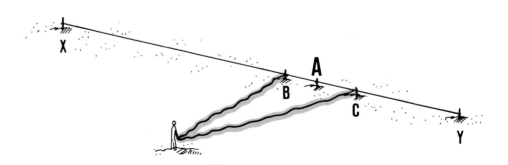

Loop the ends of the line over stakes B and C

131

19. Taking the line by its mid-point D, **stretch the line tightly**, making sure that the two ends are still looped over stakes B and C. Mark the position of the **mid-point** D with a stake. Line DA will be perpendicular to line XY.

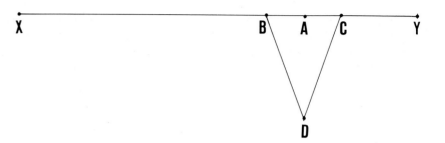

Stretch the line tight from its mid-point to find point D

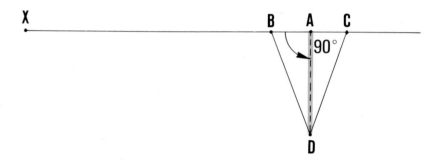

Connect points A and D to form the perpendicular

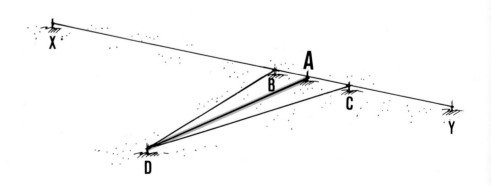

132

Setting out a perpendicular by the intersection method

20. To set out a perpendicular by the intersection method, you can use a simple line again. The method you use will depend on the length of the line. Remember that:

- if the perpendicular is to be short, it is best to use the first method (steps 21-29);
- if the perpendicular is to be long, it is best to use the second method (steps 30-38).

Using the short-line intersection method

21.   To use this method, you will need a **simple measuring line** such as a liana or a rope **5 to 6 m long**, a short pointed stick or thin piece of metal (such as a big nail) and five marking stakes.

22.   Set out line XY. On this line, choose point A, from which you will set out the perpendicular, and mark it clearly with a stake.

23.   With part of your measuring line, measure a **2 to 3 m distance to the left** of point A on XY. Mark this point B with a stake.

24.   Measure the **same distance** on XY to the right of point A. Mark this point C with a stake.

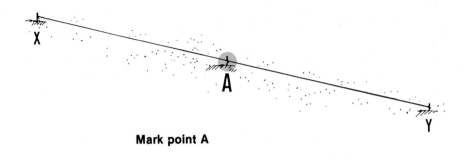

Mark point A

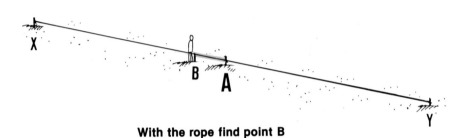

With the rope find point B

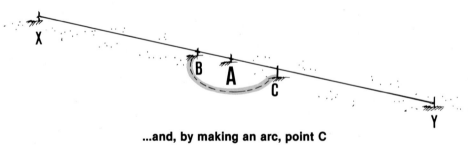

...and, by making an arc, point C

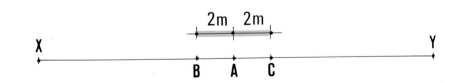

25. **Make a fixed loop** at one end of your line, and securely attach **the pointed stick** or **piece of metal** to the other end.

26. Place this loop around marking stake B. Then, **keeping the line well-stretched**, trace a large **arc** on the ground with the other end of the line. This arc should extend beyond point A, and a long way on either side of XY.

27. Take up the loop from stake B and place it around stake C. Trace another arc on the ground **which should intersect the first arc** at two points, D and E.

28. Clearly mark these two points D and E with stakes.

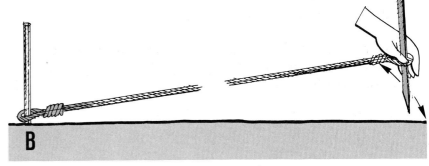

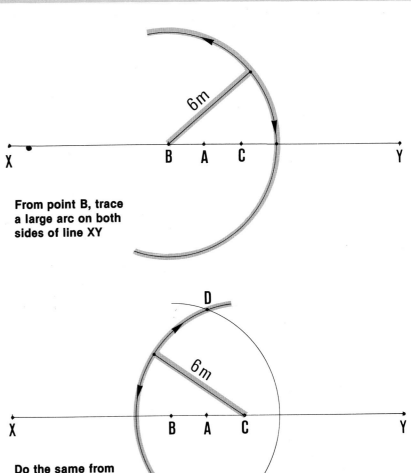

From point B, trace a large arc on both sides of line XY

Do the same from point C - the arcs intersect at points D and E

135

29. Taking up the loop from stake C, place it around stake D; holding the other end of the line, walk to stake E and attach it there; check to see if the line touches the central stake A (remember that the perpendicular was originally to be set out from point A); if it does, line DE forms this perpendicular on the ground.

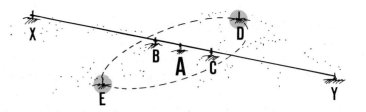

Attach the line at point D and walk to point E

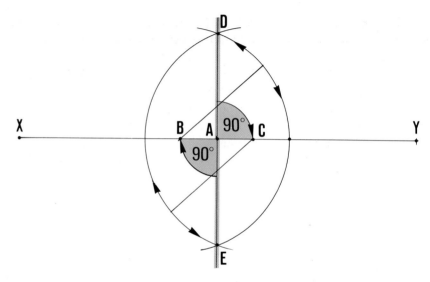

The stretched line forms perpendicular DAE

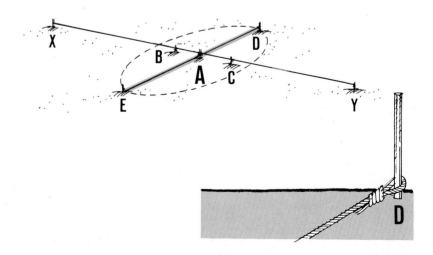

Using the long-line intersection method

30. To use this method, you will need **a simple line about 55 m long**, a short pointed stick or thin piece of metal and four marking stakes.

31. Clearly mark point A on line XY with a stake. You will set out the perpendicular from this point.

32. Measure **25 to 30 m** to the left of point A on line XY, using part of your measuring line; mark this point B with a stake.

33. Measure the same distance on XY to the right of A; mark this point C with a stake.

34. **Make a fixed loop** at one end of your line. Securely attach **the pointed stick or piece of metal** to the other end of the line (as in step 25, above).

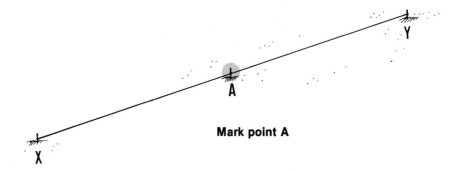

Mark point A

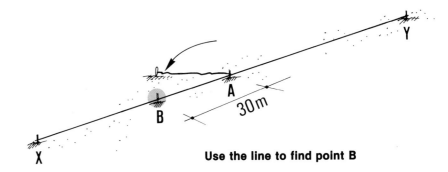

Use the line to find point B

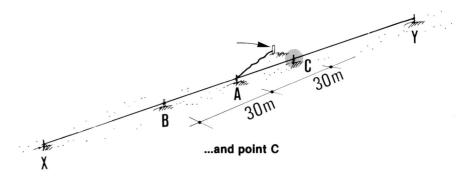

...and point C

35. Place the loop around **marking stake** B and, with the other end of the line in one hand, walk diagonally away from line XY. When you reach a point above A **where the line is well stretched**, trace an arc 2 to 3 m long on the ground with the end of your line.

36. Repeat the last step from the second stake C. The arc you mark on the ground from this point should intersect the first arc at point D.

37. At this intersection point D, drive a marking stake into the ground.

38. The line AD joining D with the original point A is perpendicular to XY.

Note: you can only use the intersection method on ground that is clear of large rocks and high vegetation, because you must be able to mark and see the arcs easily. If necessary, you can clear the ground as you work.

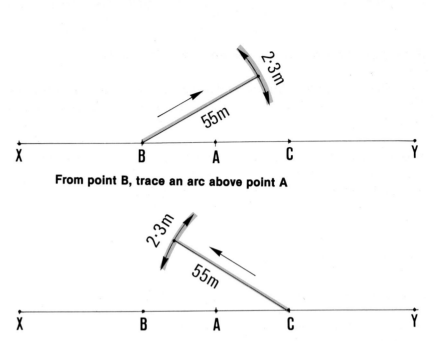

From point B, trace an arc above point A

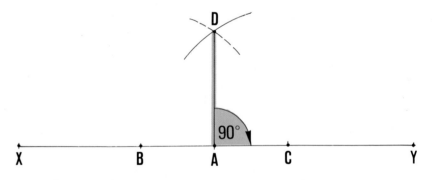

Do the same from point C - the arcs intersect at point D

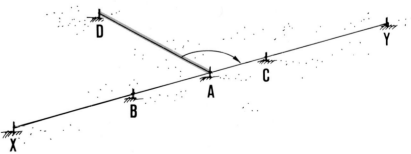

Connect points D and A to form the perpendicular

Setting out a perpendicular by the 3:4:5 rule method

39.   The 3:4:5 rule is that any triangle with sides in the proportion 3:4:5 has a right angle opposite the longest side. The method is based on this rule.

The length of the simple line you use for measuring will depend on the length of the perpendicular you are setting out. The longer the perpendicular, the longer your **measuring line** must be.

Examples

Very short line: about 1.5 m long, a little longer than 0.3 m + 0.4 m + 0.5 m = 1.2 m; **Short line**: about 13 m long, a little longer than 3 m + 4 m + 5 m = 12 m;
Medium line: about 38 m long, a little longer than 9 m + 12 m + 15 m = 36 m;
Long line: about 65 m long, a little longer than 15 m + 20 m + 25 m = 60 m.

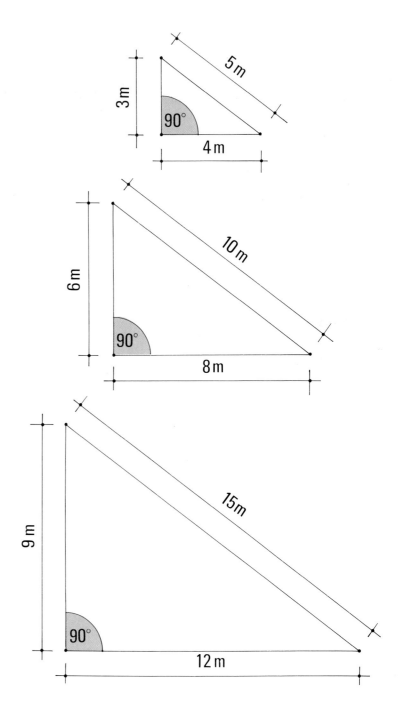

40.   To make your **simple line**, get a rope 1-1.5 cm thick; it is best if made from natural fibres which will stretch or shrink very little. A piece of **used** sisal rope will stretch or shrink less than a new one. You can also use a measuring tape.

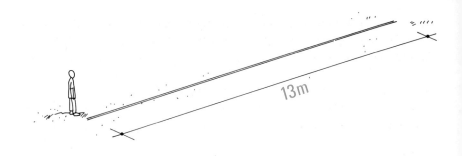

41.   There are **several ways** of using the 3:4:5 rule method, depending on the type of measuring line you use and the number of people who can work together with you. When using medium or long lines, it is best to work in a team of three people. When using a short or very short line, you can work by yourself.

Making your own 3:4:5 measuring line

Tie the ring on securely

42.   You can easily make a **simple line** to use with the 3:4:5 rule method. This line is sometimes known as **a ratio rope**. The following shows one way of making a short line about 13 m long, but you can make shorter or longer lines in the same way.

43.   Take a piece of rope about 13 m long. A few centimetres from one end, tie a metal ring to it securely with heavy string.

44.   Measure a length of 3 m along the rope from this ring, and attach a second ring to the rope.

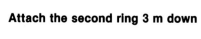

45.   Using a measuring tape, check that the distance from the first to the second ring is exactly 3 m. If it is not, adjust it.

Attach the second ring 3 m down

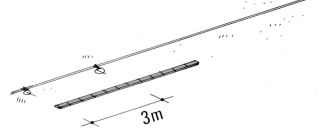

Measure again when the ring is in place

46.   Measure a length of 4 m from the second ring and attach a third ring. With a measuring tape, check that the distance is exactly 4 m. Adjust it if necessary.

47.   Measure a length of 5 m from the third ring. Attach this end point of the rope to the first ring. Check the length with a measuring tape and adjust if necessary.

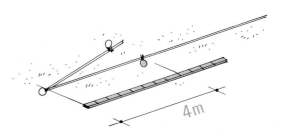

Attach the third ring 4 m down and check the measurement

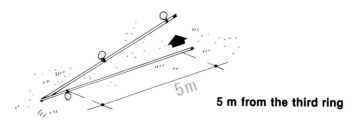

5 m from the third ring

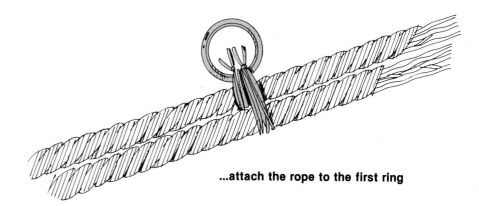

...attach the rope to the first ring

Check the final measurements

Using the short 3:4:5 line to set out a right angle

48.   Set out straight line XY on which you want to construct the right angle using a short line. Get several wooden or metal stakes.

49.   Pin or stake the ring between the 3 m and 4 m segments of the short ratio rope at point A on XY. This point could be the corner of a rectangular pond you want to build.

50.   Stretch the 3 m segment tightly along line XY, and pin or stake it at point B, using the ring between the 3 m and 5 m segments.

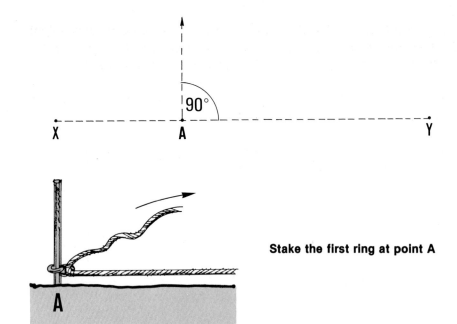

Stake the first ring at point A

Stretch the 3 m segment to point B

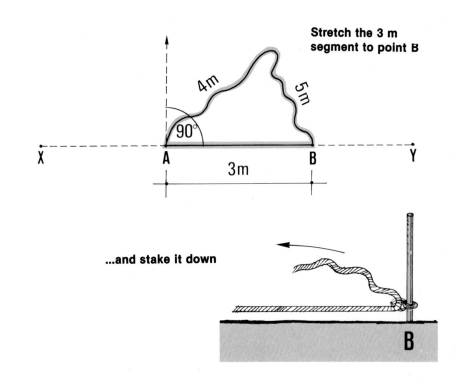

...and stake it down

142

51.   Hold the ring between the 4 m and 5 m segments, and pull the ratio rope into the shape of a triangle, making sure that the 4 m and 5 m segments are **tightly stretched**. Using this ring, stake the rope at C.

52.   The angle which has been formed at A between the 3 m and 4 m segments of the ratio rope is a right angle.

Note: you can also use a 3:4:5 ratio rope with much shorter segments. A 30 **cm: 40 cm: 50 cm line**, for example, is ideal for **measuring angles of smaller areas**, which can be used for laying out a V-knotch weir, for example (see Vol.4, **Water**, Section 36, page 72).

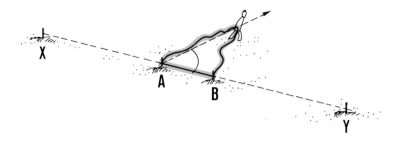

With the third ring, walk away from line XY

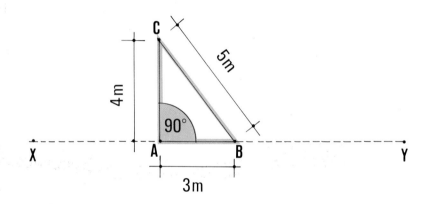

**When the rope is stretched tight,
mark point C to form the right angle**

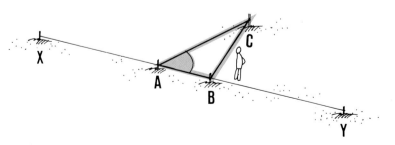

Using the medium 3:4:5 line to set out a right angle

53.   Use a **line about 36 m long**, prepared like the short line, except that the sections should be 9 m, 12 m and 15 m long. Starting at point A, where the right angle has to be set out, stretch the 12 m segment along line XY; at this point attach the ring on the line to stake B.

54.   With the 15 m segment, walk away from B while your assistant returns to the original point A with the 9 m segment of the line.

55.   When these last two sides of the triangle are **fully stretched**, mark the point C which connects the 9 m and 15 m lengths. This point forms perpendicular AC at A.

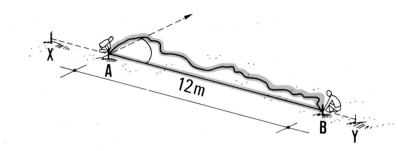

With a longer line, use an assistant

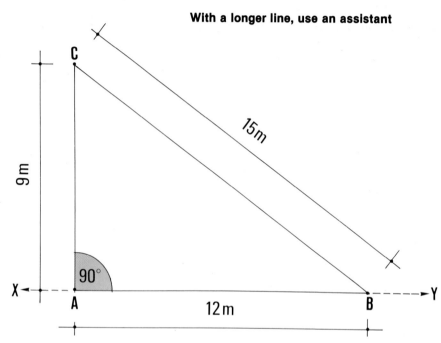

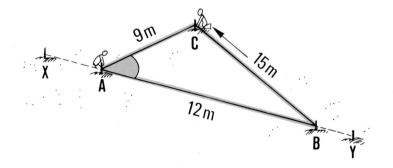

144

Using the long 3:4:5 line to set out a right angle

56.   On a **rope about 65 m long**, clearly mark the 0 m, 15 m, 35 m and 60 m lengths. You should work in a team of three people when using this line.

57.   The **first person** holds the rope at the 15 m mark at point B on line XY, from which the perpendicular is to be set out.

58.   The **second person** holds the 0 m and 60 m marks together at point A along XY.

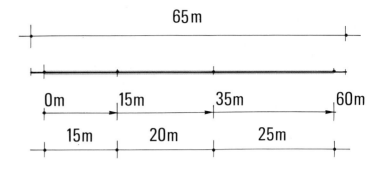

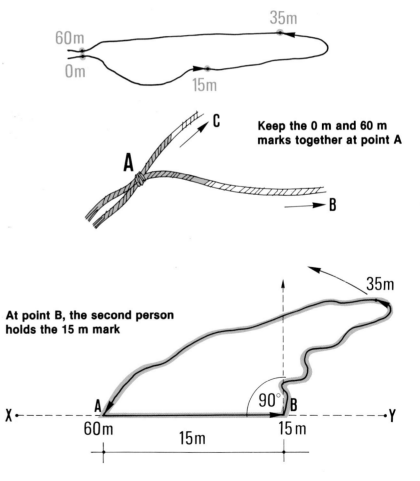

Keep the 0 m and 60 m marks together at point A

At point B, the second person holds the 15 m mark

145

59. The **third person** takes the rope at the 35 m mark and walks away from XY. He or she adjusts his or her position until the two sides of the triangle are stretched. When this is done, the position point C is marked. Joined to point B, this forms perpendicular BC to XY.

Note: you should always double-check distances to make sure that no errors have been made.

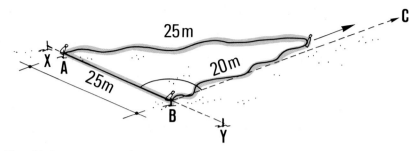

The third person walks away, holding the 35 m mark

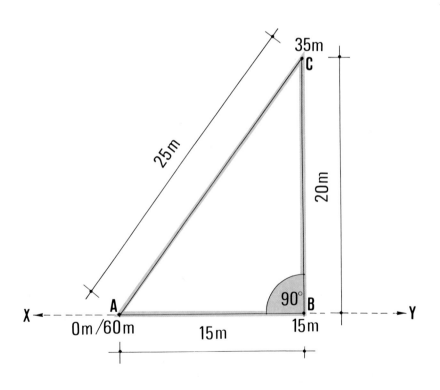

**When the rope is stretched tight,
mark point C to make the right angle**

Using a measuring tape to set out a right angle

You need to set out the centre-line WZ of a dike perpendicular to the centre-line XY of another dike, for example. Using **a measuring tape at least 80 m long** and working in a team of three, proceed as follows:

60.   From the intersection point A of the two dike centre-lines, measure 40 m along XY, the known centre-line. Mark this point B.

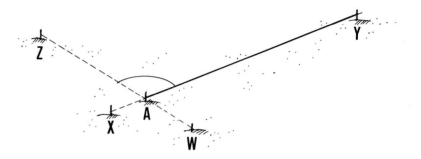

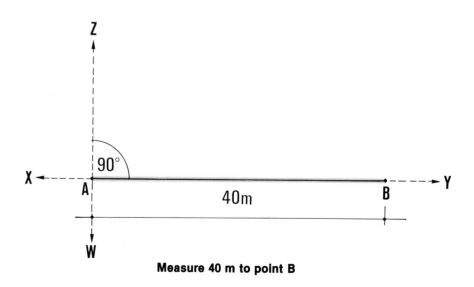

Measure 40 m to point B

61. **While one person** holds the 0 m graduation of the tape at point B, the **second person** holds the 80 m graduation at point A, where the two centre-lines intersect.

62. The **third person**, holding the tape at the 50 m graduation, walks away from XY until the tape is fully stretched. He or she clearly marks the place where he or she stands point C. This point defines the second centre-line WZ perpendicular to the first one.

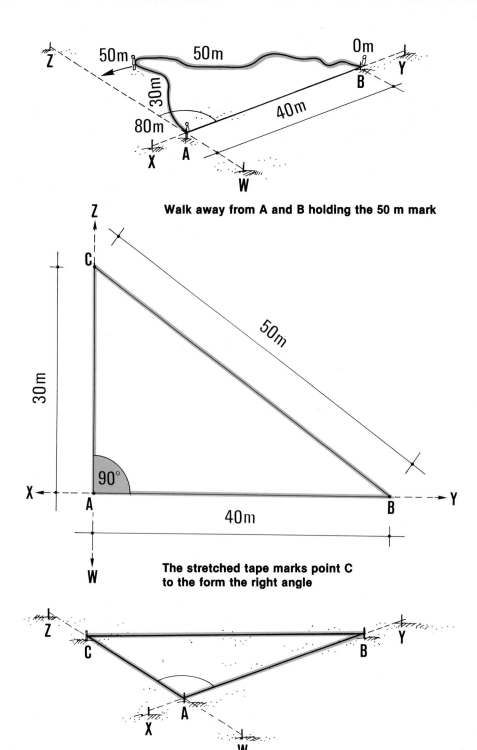

Walk away from A and B holding the 50 m mark

The stretched tape marks point C to the form the right angle

148

Setting out a perpendicular with a cross-staff

63.   **A cross-staff** is an inexpensive sighting instrument which is very useful for setting out right angles. There are several models, such as the **octagonal brass cross-staff**, which has sighting slits cut at right angles to each other, and the **foresight/backsight model**. In use, cross-staffs must be firmly fastened to a **support**, usually a stake driven vertically into the ground. Their useful range does not extend beyond 30 to 40 m. You might be able to borrow a cross-staff from a surveyor's office, or you can build your own as described below.

Note:the **octagonal cross-staff** also has additional **sighting slits cut at 45°**, and is useful for setting out 45-degree angles (see, for example, Section 29, step 7).

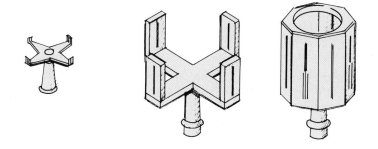

Professional cross-staffs

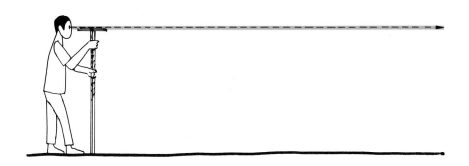

Sighting with a home-made cross-staff

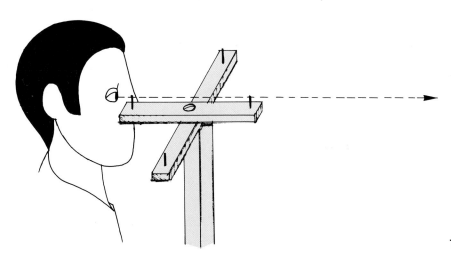

Making your own cross-staff

64.   Get two metal or wooden strips 2 to 3 cm wide and 20 to 25 cm long. Find the centres of the strips at the intersection of two diagonal lines, as you found the centre of your alidade in step 9. Drill a small hole exactly at this centre point you have found on each strip. These are the **cross-pieces**.

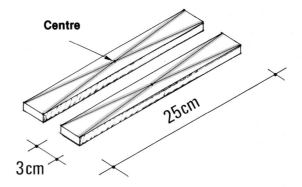

65.   Provide a **sighting line** on each cross-piece. To do this on wooden strips, drive in one small headless nail, centred, near each end of each strip. On metal strips, you can weld or glue small nails or metal points near the ends of the strips.

66.   Place the cross-pieces approximately at right angles and, with a screw, attach them loosely in that position to the top of a 1.50 m vertical stake. If you use washers between the wooden cross-pieces and the stake, it will be easier to tighten the strips securely later.

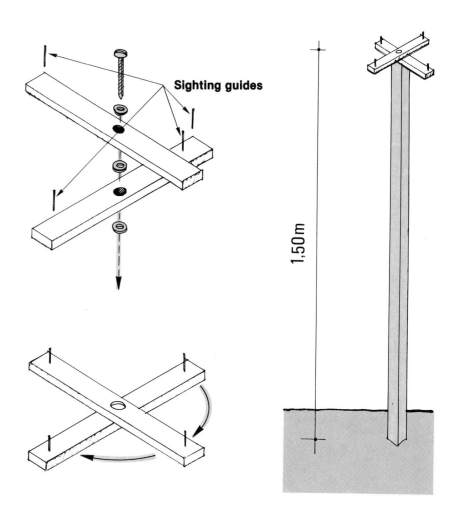

150

Adjusting the home-made cross-staff

67.   **Lay out a right angle on the ground** using a long 3:4:5 ratio rope (see steps 56 to 59, this section). The sides of the triangle will be 15 m, 20 m and 25 m long.

68.   Put a short stake at point A, the corner of the right angle, between the 15 m and 20 m sides. Then put ranging poles at points B and C to mark the sides of the angle.

69.   Position the cross-staff and its vertical support at point A.

70.   **Align one cross-piece** along side AB and sight towards point B.

71.   Without moving the vertical support, **align the second cross-piece** along the other side AC of the angle, and sight towards point C. Tighten the screws a little to keep the cross-pieces in place.

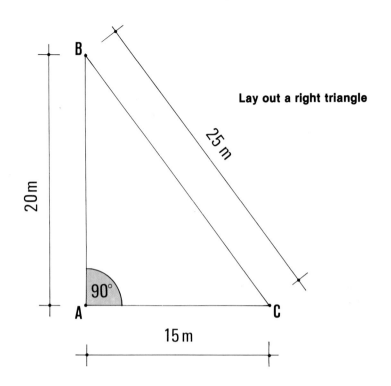

Lay out a right triangle

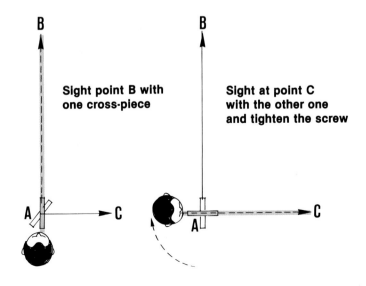

Sight point B with one cross-piece

Sight at point C with the other one and tighten the screw

72.   **Rotate the vertical support 90°** to check that the cross-pieces are truly at a right angle. Sight at points B and C again and correct the position of the cross-pieces if necessary.

73.   Repeat this process until you are sure that **each cross-piece is aligned with one side of the right angle**, that is, that they are at 90° angles from each other.

74.   When both cross-pieces are properly aligned, **firmly tighten the screw** holding them to the vertical support.

75.   Check both sighting lines again after tightening the screw to make sure that the cross-pieces have not slipped.

76.   To help you adjust the cross-pieces later, **cut or engrave** (with a large nail) **marks** in the wood or metal of the bottom cross-piece when the top piece is in position.

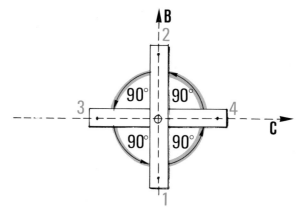

Rotate the cross-staff to check its accuracy

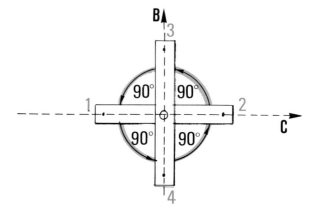

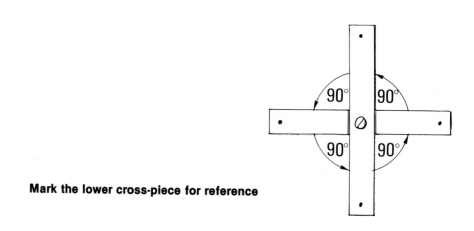

Mark the lower cross-piece for reference

Using the cross-staff to set out a right angle

77. To use the cross-staff, you will need an **assistant**.

78. Lay out straight line XY on which you need to construct the right angle at point A.

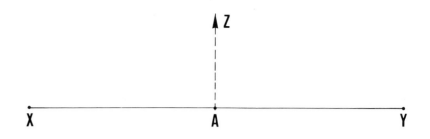

79. Place the support of the cross-staff in a vertical position at point A.

80. Ask your assistant to hold a ranging pole in a vertical position at point B, near the end of XY.

81. Sight along one of the cross-pieces and rotate the vertical support until the sighting line is aligned on B.

Sight along XY to align the cross-staff

82. Without moving the cross-staff and its vertical support, sight along the other cross-piece. At the same time, direct your assistant to stand with a ranging pole as near to this sighting line as possible.

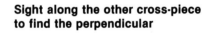

**Sight along the other cross-piece
to find the perpendicular**

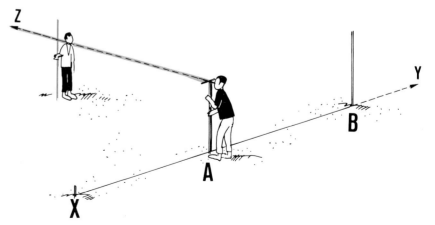

83. Tell your assistant to hold the ranging pole in front of him or her and move to the right or left until the pole is exactly on the sighting line AZ.

84. When you are sure he or she is on line AZ, tell him or her to mark his position with stake C.

85. The angle BAC you have determined at point A, where the cross-staff is placed, is a 90° angle.

Note: with the help of a cross-staff you can easily determine the rectangular areas which you need for a fish-pond layout. You can also build a grid of squares by determining intermediate angles along your straight lines. This is a method used in estimating reservoir volumes, for example (Volume 4, **Water**, Section 42, page 107).

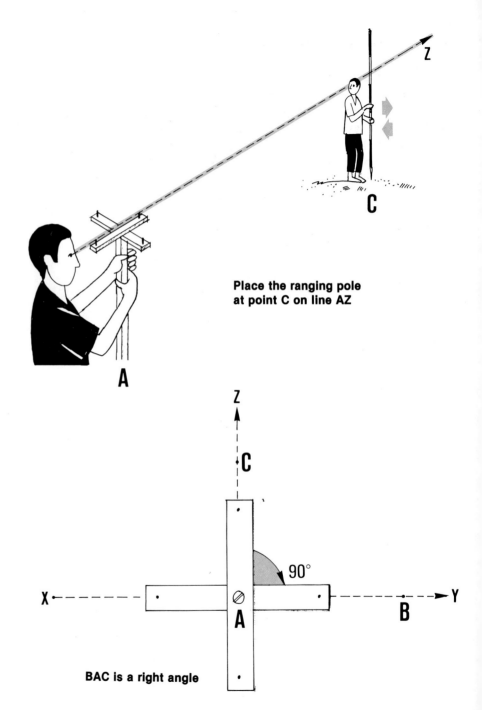

Place the ranging pole at point C on line AZ

BAC is a right angle

37   How to set out parallel lines

What are parallel lines?

1.   Parallel lines, also called parallels, are **lines equally distant from each other at every point**. Parallel lines run side by side and will **never** cross. They are very important in fish culture and are often used in designing fish-farms (for example, for parallel dikes and ponds), in building dams and in setting out water canals. Parallels are also useful when laying out lines under difficult conditions (see Section 30).

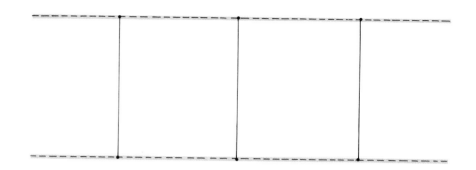

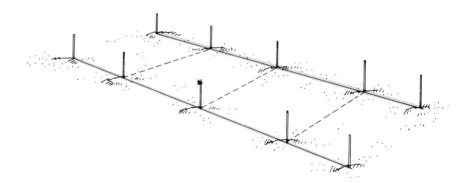

Setting out parallels by the 3:4:5 rule

One way to set out a parallel line uses the 3:4:5 rule. It works like this:

2. On given line XY, **select two points** A and B which are fairly distant from each other (for example, 20 to 30 m apart), and mark them with pegs.

3. From each of these points, **set out a perpendicular*** using the 3:4:5 rule method. **Remember** that the length of the line you will use depends on the length of the perpendicular you are setting out (see Section 36, step 35).

4. Prolong these two perpendiculars as required. Then, measure **an equal distance** from the given line XY on each of them; mark these two points C and D.

5. Through these two points, set out a line WZ. This line will be parallel to XY.

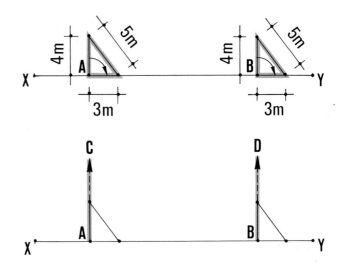

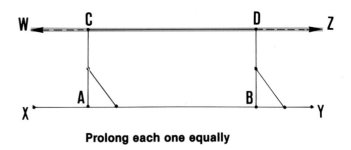

Set out two perpendiculars

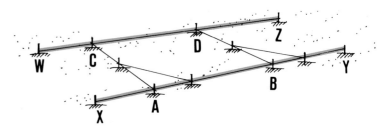

Prolong each one equally

Connect the points to form the parallel

Setting out parallels with the crossing-lines method

To use the crossing-lines method, you do not need to set out perpendiculars; you will only measure distances. However, this method cannot be used when you need to measure the exact position of **the parallel** you need to set out. It is useful when the distance of the parallel is not important, such as when you are prolonging a line over an obstacle (see Section 30). Proceed as follows:

6.   Lay out line XY. **Select any point A which will belong to the parallel line** you need to set out. Clearly mark point A with a peg.

7.   From point A, set out an **oblique line** AZ. Mark point B where AZ intersects the original line XY.

Note: an oblique line is neither parallel, nor perpendicular.

8.   Measure the **length** of the oblique line AB.

9.   Divide this length by 2, measure this distance from point A, and mark this **central point C**.

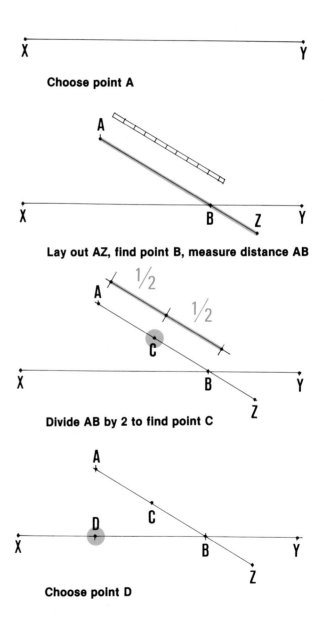

Choose point A

Lay out AZ, find point B, measure distance AB

Divide AB by 2 to find point C

Choose point D

157

10. On the original line XY, select point D, which should lie as **nearly opposite** point A as possible.

11. From point D, set out a straight line DW **passing through point C**.

12. Measure distance DC.

13. From point C on line DW, measure a distance equal to distance DC. Mark the end of the segment point E.

14. Connect points E and A with **a line KL.** This line is **parallel to line XY**.

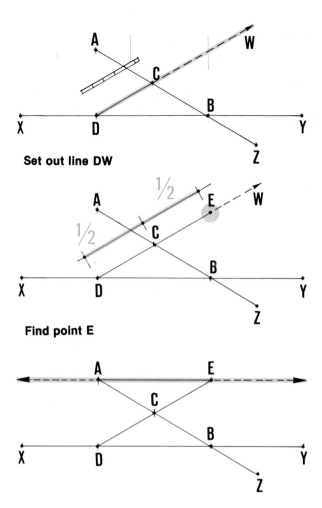

Set out line DW

Find point E

Connect the points A and E to form the parallel

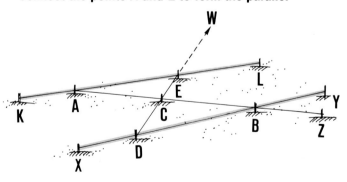

Setting out a series of rectangular areas

15. When you build a fish-farm, you will often need to set out a series of rectangular plots on the ground. These plots are the future sites of ponds or other constructions, (see the next manuals in this series, **Constructions for Freshwater Fish Culture)**.

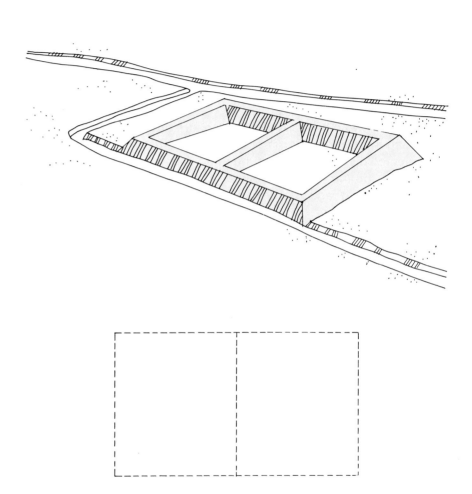

16. First select the direction of the **main dike's centre-line XY**, and set it out on the ground with ranging poles. Using the measurements along this line, you will be able to mark the points A, B and C where you will set out the centre-lines of the secondary dikes. To do this, proceed as follows:

17. Set out a few **perpendiculars*** on XY, using one of the methods given in Section 36, for example, from two **extreme points** A and B (near the end-points of XY) and from one **intermediate point** C.

18. Starting from points A and B, measure **equal distances** AF and BG along their perpendiculars; these distances should be equal to the selected distance between the main dike centre-line XY and the centre-line of the opposite dikes. Mark the two points on the perpendiculars, F and G, with ranging poles.

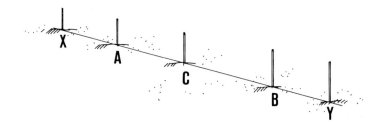

**Set out the centre line of the main dike
and the points where the other dikes will cross it**

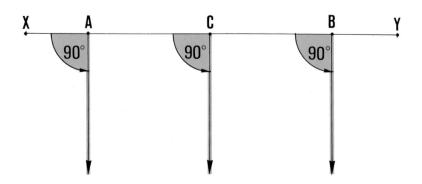

Lay out perpendiculars from these points

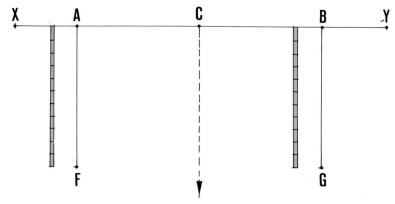

Measure equal distances along the perpendiculars

19. Clearly set out line WZ through points F and G using ranging poles.

20. Starting from point B on line XY, measure **intermediate distances** BE, EC and CD. Then move back to line WZ; starting from point G, measure **intermediate distances** GH, HI and IJ equal to BE, EC and CD, respectively. Mark points H, I and J with pegs.

21. While you are doing this, **check that point I falls exactly on the intermediate perpendicular set out from C**. If there is a small difference, adjust the positions of the perpendicular and point I. If there is a large difference, check your previous work for errors.

22. As a final check, be sure that the last measurement JF lines up with point F.

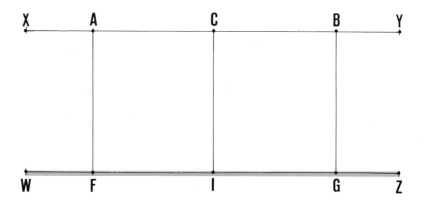

Set out parallel WZ

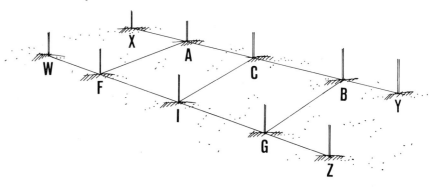

Set out and check intermediate distances

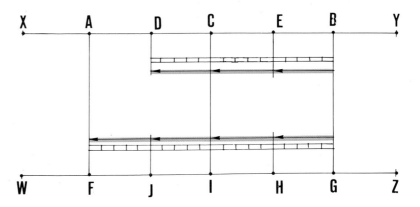

161

4   MEASURING VERTICAL ANGLES AND SLOPES

1.   A **vertical angle** is an angle formed by two connected lines in **the vertical plane***, that is, between a low point and two higher points. Since these angles are in the vertical plane, the lines that form them will usually be lines of sight. **A vertical angle** BAC can be formed, for example, by the line of sight AB from station A on a river bank to a higher water-pump installation, and the line of sight AC from station A to a much higher water-storage tank.

2.   Whenever a line is not horizontal, it has a **slope**. The slope can be uphill or downhill. Its steepness depends on the difference in height between its points.

3.   As you have learned (see Chapter 2), the slope of the ground affects **the measurement of distances**. Ground slope is also very important in the design of fish-farms, since you can use it to reduce your **construction costs**. You need to build bottom slopes in canals, to allow the water to move by **gravity***; and in ponds, to allow good drainage. And you must build slopes in the dikes for ponds and dams (see the next manuals in this series, **Constructions for Freshwater Fish Culture**).

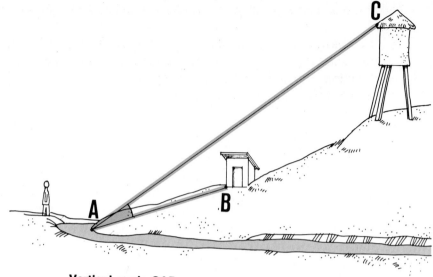

Vertical angle CAB

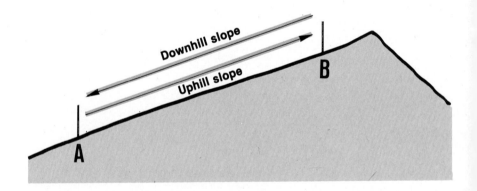

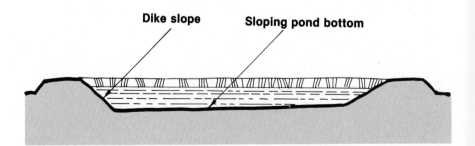

Dike slope   Sloping pond bottom

4. The slope of a line is called the **gradient**. It may be defined as:

- **The change in vertical distance or elevation*** over a given horizontal distance, or the **change in horizontal distance or elevation** over a given vertical distance;
- **The vertical angle** made by the sloping line and a horizontal line.

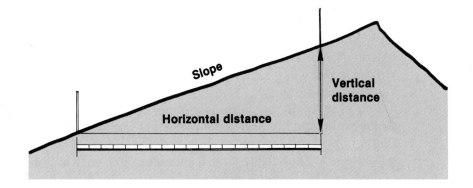

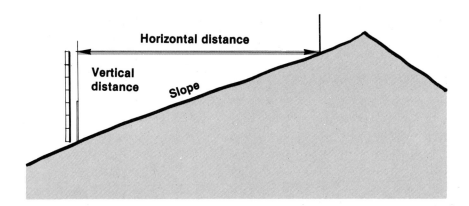

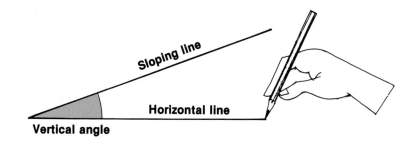

5. The slope of a line is therefore expressed in various ways:

- as a percentage, or the number of metres of change in elevation over a horizontal distance of 100 m. This may be written in two ways, either as a percent (%) or as a decimal value, in hundredths;
- in degrees, as the measurement of the vertical angle made by the slope and the **horizontal plane***.
  Remember that:
  - degrees are subdivided into 60 **minutes** (60′), each minute equalling 60 **seconds** (60″);
  - a right angle equals 90°, and therefore a slope is always measured between 0° (horizontal) and 90° (vertical);

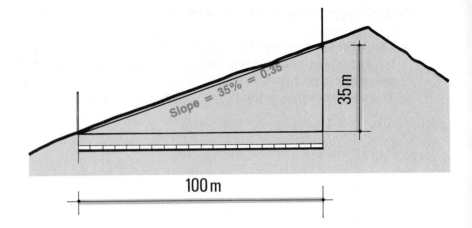

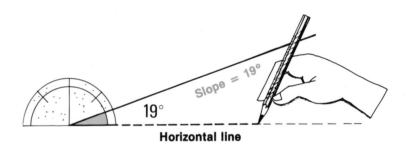

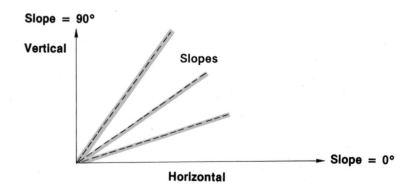

166

- As a ratio, showing the change in horizontal distance (x) per unit of vertical distance, or the change in vertical distance (y) per unit of horizontal distance, in one of the following ways:
  - the change in horizontal distance (x metres) per one metre of vertical distance; this can express, for example, the slope of the sides of dikes and canals (such as 2:1);
  - the change in vertical distance (x millimetres or x centimetres) per one metre of horizontal distance; this can express, for example, the lengthwise slope of a pond bottom or water pipe (such as 3 cm/m);
  - the change in horizontal distance (x units) per one unit of vertical distance. This can express, for example, the lengthwise slope of a pipeline (such as 1 in 300).

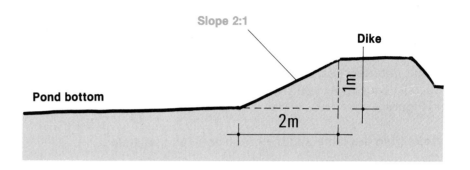

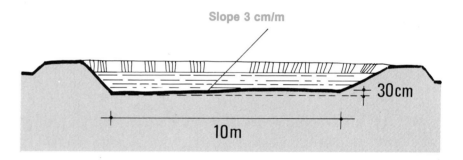

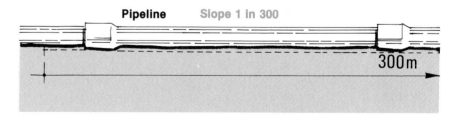

1 m change in elevation every 300 m

Converting percentage of a slope into degrees, or degrees into percentage

6.   Depending on the instrument you are using to measure a slope directly, you may sometimes have **to convert the percentage** of the slope into degrees, or the degrees into percentage. For help with such a conversion, you should use either **Table 4** or the graph given in **Figure 3**.

Note: from the table and the graph you can see that:

- **1 degree** is about **1.75 percent**;
- **1 percent is about 0º35′**;
- A **45º slope** = a **100 percent slope**.

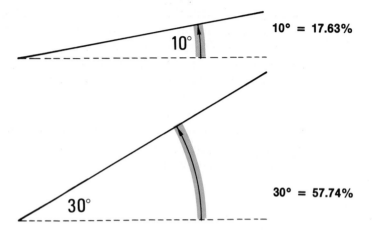

10° = 17.63%

30° = 57.74%

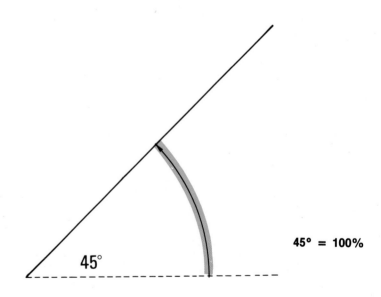

45° = 100%

TABLE 4

Conversion of slope units in degrees or percentages

From percent into degrees		From degrees into percent			
Percent	Degrees/min/sec	Degrees	Percent	Degrees	Percent
0.5	0°17'10"	0.25(15')	0.44	11	19.44
1	0°35'	0.50(30')	0.87	12	21.26
2	1°08'40"	0.75(45')	1.31	13	23.09
5	2°51'40"	1	1.75	14	24.93
10	5°42'40"	2	3.49	15	26.79
20	11°18'36"	3	5.24	16	28.68
30	16°42'	4	6.99	17	30.57
40	21°48'05"	5	8.75	18	32.49
50	26°33'55"	6	10.51	19	34.43
100	45°	7	12.28	20	36.40
		8	14.05	30	57.74
		9	15.84	40	83.91
		10	17.63	45	100

Remember: 60 min = 1 degree and 60 sec = 1 min

Examples:

- a slope of 17 percent is equal to (10 + 5 + 2) percent, which is equivalent to 5°42'40" + 2°51'40" = 8°101'120" = 8°103' = 9°43';

- a slope of 9°43' is about equal to (9° + 30' + 15'), which is equivalent to 15.84 percent + 0.87 percent + 0.44 percent = 17.15 percent or 17 percent.

FIGURE 3

Graph for the rapid conversion of slope units

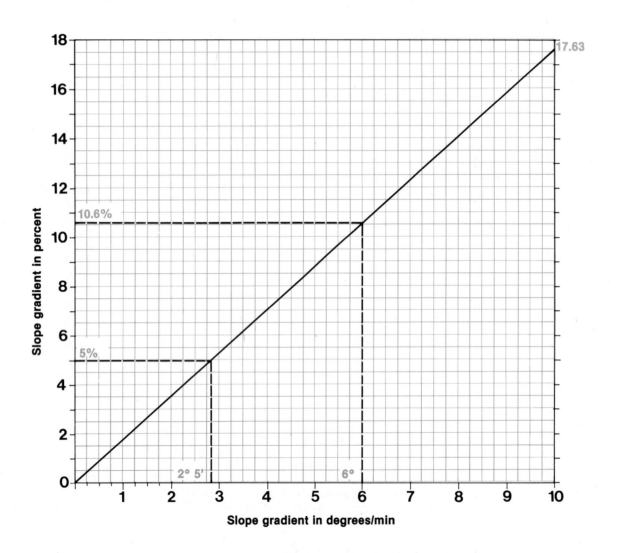

Measuring and calculating slopes?

7. There are two groups of methods for determining slopes:
    * You can measure the slope directly, using one of the devices described later in this chapter (see step 14). In this case you read the gradient (in degrees or in percent) from the instrument, without making any further calculations; or
    * You can calculate the slope: measure the ground-level difference (in metres) between two points along the steepest part of the slope (called the axis), using one of the devices described in Chapter 5. Calculate the slope, which you will usually express as a percentage (see next step).

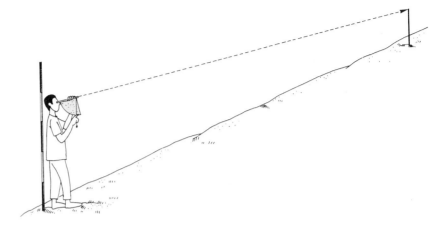

Measuring slope directly

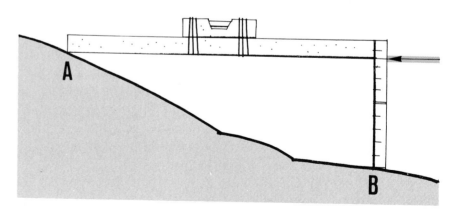

Measuring ground-level differences to calculate slope

8. To calculate the slope, proceed as follows:

- along the axis of the slope, measure the **difference in level AC** between two marked points A and B (see Chapter 5);
- measure the **horizontal distance CB** between points A and B (see Chapter 2);
- calculate the **slope S** in percent as equal to:

$$S\% = 100\ AC - CB$$

Note: to make your calculations easier:

- you can **fix the horizontal distance CB at 100 m**, which will give you S% = **AC** directly in metres;
- you can **fix the horizontal distance CB at 10 m** instead, which will give you S% = 10 **AC** in metres.

Remember: you must measure the **horizontal distance**!

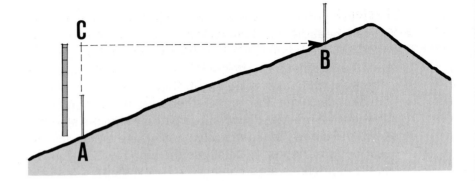

First measure the difference in level

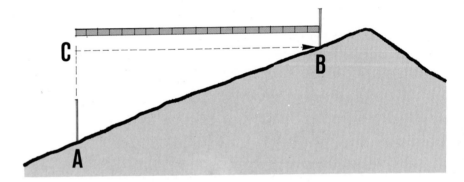

Then measure the horizontal difference

Using slope to calculate horizontal distances

9. In Chapter 2, Sections 26 and 27, you learned that when measuring a distance AB on sloping ground, you need to correct this measurement in order to find the **true horizontal distance AC**, but only when the slope **exceeds 5 percent (or about 3 degrees)**. To make these corrections, you may use either the method described below, or the method which will be described in **Section 50, step 17**. To calculate horizontal distances from distances measured over sloping ground, proceed as follows:

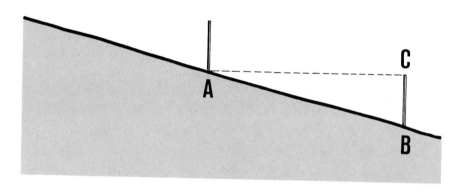

10. Measure the **distance AB (in metres) on the ground** between points A and B (see Chapter 2).

11. Measure **the average ground slope S in degrees** between points A and B (see this chapter, Sections 41 to 47).

Note: if the slope is measured in percent, you will have to convert it into degrees (see Table 4 or Figure 3).

12. Enter this average slope **S (in degrees)** in **Table 5** to obtain the value of **cosine S** (cos S). If the slope does not correspond exactly to any of the angle values given in the table, you will have to calculate cos S by using **proportional parts** (see example in Table 5).

13. Calculate the **horizontal distance AC** (in metres) using the formula:

$$AC = AB \cos S$$

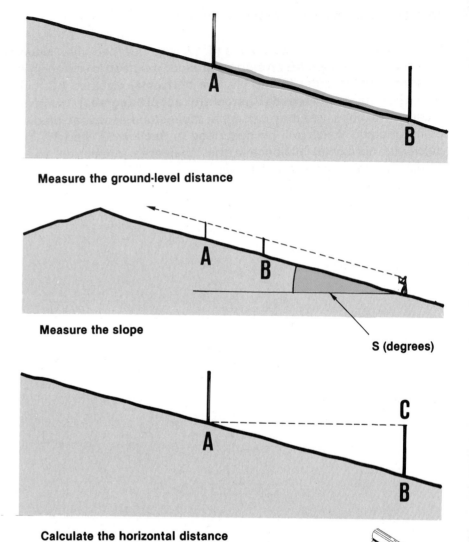

Measure the ground-level distance

Measure the slope

S (degrees)

Calculate the horizontal distance

AC = AB cos S

174

TABLE 5

Cosine values of angles

(*d* = degrees, *m* = minutes, *cos* = cosine, *x* = difference)

MAIN TABLE

d	m	cos	x
3	0	0.9986	
	10	0.9985	1
	20	0.9983	2
	30	0.9981	3
	40	0.9980	1
	50	0.9978	2
			2
4	0	0.9976	
	10	0.9974	2
	20	0.9971	3
	30	0.9969	2
	40	0.9967	2
	50	0.9964	3
			2
5	0	0.9962	
	10	0.9959	3
	20	0.9957	2
	30	0.9954	3
	40	0.9951	3
	50	0.9948	3
			3
6	0	0.9945	
	10	0.9942	3
	20	0.9939	3
	30	0.9936	3
	40	0.9932	4
	50	0.9929	3
			4
7	0	0.9925	
	10	0.9922	3
	20	0.9918	4
	30	0.9914	4
	40	0.9911	3
	50	0.9907	4
			4
8	0	0.9903	
	10	0.9899	4
	20	0.9894	5
	30	0.9890	4
	40	0.9886	4
	50	0.9881	5
			4
9	0	0.9877	
	10	0.9872	5
	20	0.9868	4
	30	0.9863	5
	40	0.9858	5
	50	0.9853	5
			5
10	0	0.9848	
	10	0.9843	5
	20	0.9838	5
	30	0.9833	5
	40	0.9827	6
	50	0.9822	5
			6
11	0	0.9816	
	10	0.9811	5
	20	0.9805	6
	30	0.9799	6
	40	0.9793	6
	50	0.9787	6
			6

d	m	cos	x
12	0	0.9781	
	10	0.9775	6
	20	0.9769	6
	30	0.9763	6
	40	0.9757	6
	50	0.9750	7
			6
13	0	0.9744	
	10	0.9737	7
	20	0.9730	7
	30	0.9724	6
	40	0.9717	7
	50	0.9710	7
			7
14	0	0.9703	
	10	0.9696	7
	20	0.9689	7
	30	0.9681	8
	40	0.9674	7
	50	0.9667	7
			8
15	0	0.9659	
	10	0.9652	7
	20	0.9644	8
	30	0.9636	8
	40	0.9628	8
	50	0.9621	7
			8
16	0	0.9613	
	10	0.9605	8
	20	0.9596	9
	30	0.9588	8
	40	0.9580	8
	50	0.9572	8
			9
17	0	0.9563	
	10	0.9555	8
	20	0.9546	9
	30	0.9537	9
	40	0.9528	9
	50	0.9520	8
			9
18	0	0.9511	
	10	0.9502	9
	20	0.9492	10
	30	0.9483	9
	40	0.9474	9
	50	0.9465	9
			10
19	0	0.9455	
	10	0.9446	9
	20	0.9436	10
	30	0.9426	10
	40	0.9417	9
	50	0.9407	10
			10
20	0	0.9397	

TABLE OF PROPORTIONAL PARTS, P

m	1	2	3	4	5	6	7	8	9	10	m
				Cos difference, x							
1	0.1	0.2	0.3	0.4	0.5	0.6	0.7	0.8	0.9	1.0	1
2	0.2	0.4	0.6	0.8	1.0	1.2	1.4	1.6	1.8	2.0	2
3	0.3	0.6	0.9	1.2	1.5	1.8	2.1	2.4	2.7	3.0	3
4	0.4	0.8	1.2	1.6	2.0	2.4	2.8	3.2	3.6	4.0	4
5	0.5	1.0	1.5	2.0	2.5	3.0	3.5	4.0	4.5	5.0	5
6	0.6	1.2	1.8	2.4	3.0	3.6	4.2	4.8	5.4	6.0	6
7	0.7	1.4	2.1	2.8	3.5	4.2	4.9	5.6	6.3	7.0	7
8	0 8	1.6	2.4	3.2	4.0	4.8	5.6	6.4	7.2	8.0	8
9	0.9	1.8	2.7	3.6	4.5	5.4	6.3	7.2	8.1	9.0	9

Example

To calculate intermediate cosine values using the proportional parts, for cos 7°38′ for example, proceed as follows:

- from the Main Table, calculate cos 7°30′ = 0.9914;
- obtain the difference between this value and the next, x = 3;
- find column 3 in Table of Proportional Parts, P;
- move down this column to line m = 8, to find P = 2.4;
- subtract P from the last number (4) of the value read from the Main Table, 0.9914 − 0.00024 = 0.99116. This is cos 7°38′.

Choosing a method to use for measuring slopes

14.   There are several good ways **to measure slopes**. The method you use will depend on several factors:

- how accurate a result you need;
- the equipment you have available;
- the type of terrain on which you are measuring.

Each of the various methods is fully explained and illustrated in the following sections, except for the method to use with levelling devices (see Chapter 5). **Table 6** will also help you to compare the various methods and to select the one best suited to your needs.

Clinometer

Plumb-line

Clisimeter

TABLE 6

Vertical angle and slope measurement methods

Section [1]	Method [2]	Accuracy	Remarks	Equipment [2]
41*, 42*	*Clinometer*, models 1 & 2	Low	Quick and rough estimate for rather steep slopes Hand-held instrument	Home-made clinometer
43*	*Clinometer*, model 3	Low to medium	To be fixed in ground Direct reading in percent	Home-made clinometer
44*	*Clinometer*, model 4	Low to medium	To be fixed in ground Small, easy to make Direct reading in percent	Home-made clinometer
45**	Clisimeter	Low (about 10 percent)	Quick and rough estimate Direct reading in percent	Lyra clisimeter
46**	Optical clinometer	Medium to high	Quick, rather good estimate Direct reading in degrees and percent	Optical clinometer
47***	Miscellaneous levelling devices	Medium to high	Requires distance measurement Best estimates for small gradients, especially with best levels	Various, see Table 7

[1] * Simple   ** more difficult   *** most difficult.

[2] *In italics*, equipment you can build yourself from instructions in this manual.

41   How to measure with the home-made clinometer, model 1

1.   A **clinometer** is an instrument for measuring slopes or vertical angles. There are various types of clinometers, but they all include a **graduated arc** similar to a protractor (see Section 33, step 11). To use the clinometer, you hold it in your hand and read the slope against this arc. You also usually refer to a free-hanging plumb-line called the **pendulum**. There is a **line of sight*** on the top of the clinometer. You can easily make your own simple clinometer; four models are described in Sections 41 to 44.

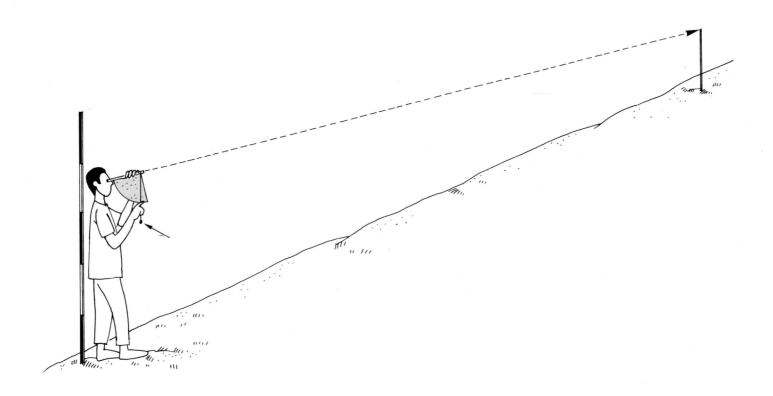

Making the pendulum clinometer, model 1

2. Get a **protractor with a 0° to 90° scale**, or make one yourself as described in Section 33. The protractor should be fairly large (for example, about 20 to 25 cm diameter) to provide reasonable accuracy.

Note: if you use **Figure 2** to make your protractor, you can easily **draw a larger 0° to 90° protractor**. To do this, get a piece of string and tie a pencil to one end. Measure 20 to 25 cm from the pencil along the string. Hold the string at this point on the centre-point A of the protractor in Figure 2. With the string stretched tightly, draw an arc with the pencil above the rounded edge of Figure 2. Then **add graduations** on your new protractor by projecting lines from the graduations in Figure 2. Glue the protractor to a piece of thin wooden board or plywood and cut carefully along its outline.

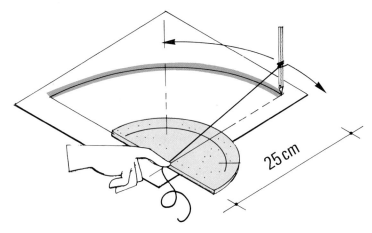

Draw a larger arc with pencil and string

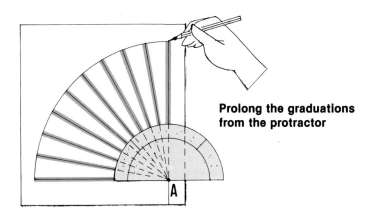

Prolong the graduations from the protractor

Glue the protractor to a wooden backing and cut it out

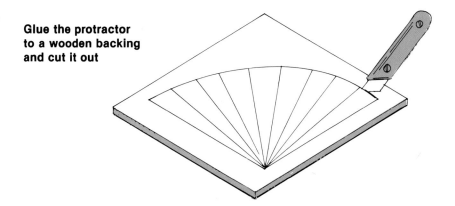

3. Attach a plumb-line (see Section 48, step 2) to a small nail driven into the centre-point A of the protractor. Make the plumb-line with a thin piece of string about 35 to 40 cm long and a small weight, such as a heavy nut or a small stone.

4. Glue **a 30-cm sighting device** along the 90° side of the protractor. To make the sighting device, get a soda straw or a narrow tube; or get a thin length of wood and attach two pins along it in a straight line.

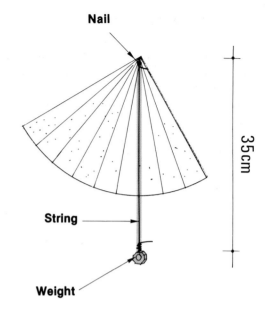

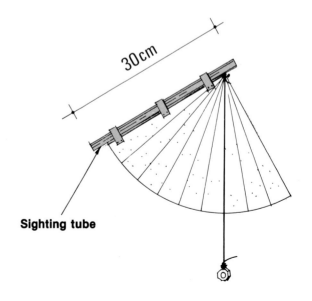

Adjusting your home-made clinometer

5. Measure the **vertical distance** from the level of your eyes to the ground, then measure the **same** vertical distance on a wall and mark it clearly. You will also need to mark this **vertical distance** clearly on a pole or staff, which you will use for sighting.

- Get **a straight pole** whose length is equal to your eye level plus 25 cm. (For example, if your eye level is 145 cm, the pole should be 145 cm + 25 cm = 170 cm.) Make a sharp point at one end and drive it into the ground until the top of the pole is at your eye level. Carefully mark the point where the pole enters the ground, called the **reference* level**; when you use the pole, always drive it in up to this line. To make the top of the pole more visible, mark it with bright-coloured paint or cloth. You will sight at the top of the pole.

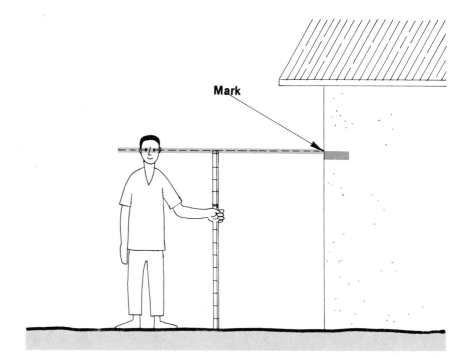

Make a mark at your eye level

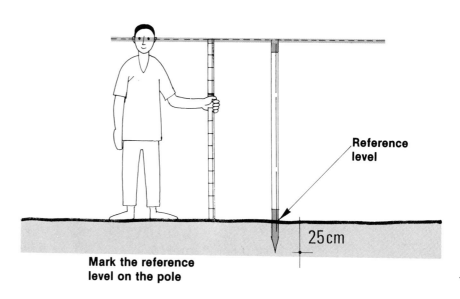

Mark the reference level on the pole

181

6. Stand on **horizontal ground** about 15 paces in front of the mark, and aim at it through the sighting device on your clinometer.

7. Check that the **plumb-line string indicates 0°**. If it does not, adjust the small nail holding the plumb-line. When the string indicates 0°, your clinometer is ready to use.

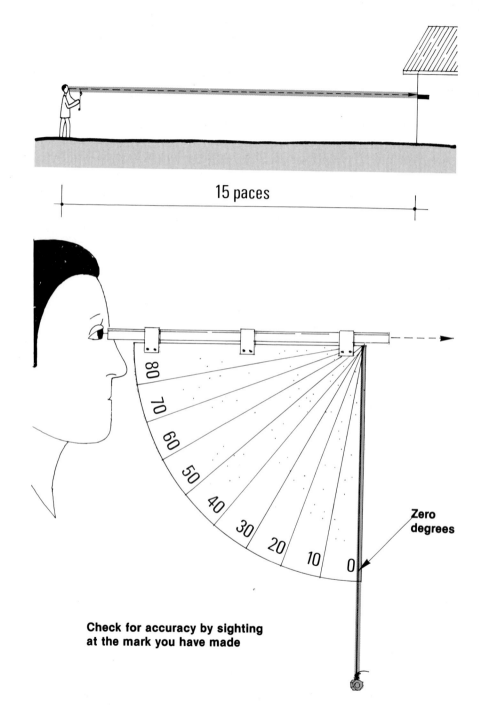

15 paces

Zero degrees

Check for accuracy by sighting at the mark you have made

Using your clinometer to measure a slope

8.   Sighting either **uphill or downhill** with the clinometer, you can measure a slope by moving the protractor around.

9.   Take a position with the clinometer. Make sure to stand up straight so you do not change your eye level. Sight at a point. This point should be:

- at eye level; use the pole or staff you prepared in step 5, making sure that it is vertical;
- no farther away than 30 m – a shorter distance (15-20 m) will improve the accuracy of the measurement.

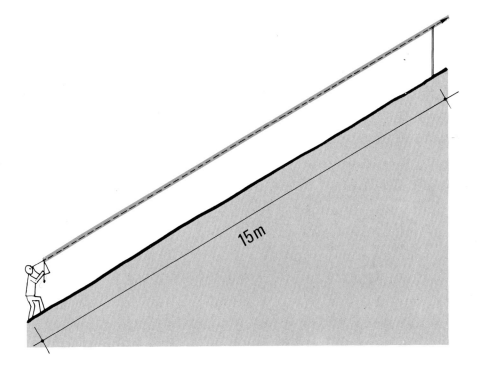

Make sure point A is at the top, whether you are sighting uphill or downhill

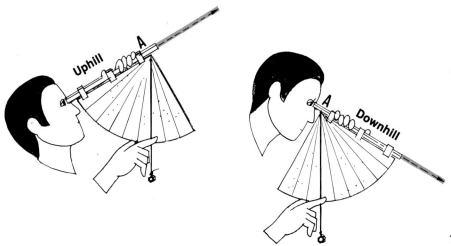

183

10.   With the clinometer in sighting position, press the plumb-line with your finger against the bottom scale. Be careful not to move the plumb-line from its vertical position. Read the scale at the point where the plumb-line intersects the degree graduation. This reading is the slope, **in degrees.**

Note: you can convert your degree measurement into a percentage (see Section 40).

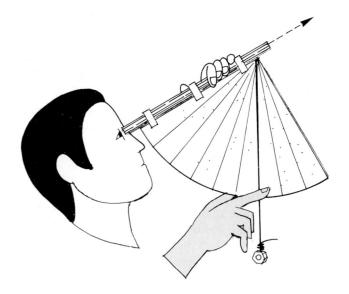

Hold the plumb-line in place with your finger

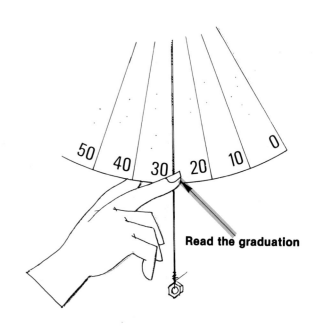

Read the graduation

42 How to measure with the home-made clinometer, model 2

1. You can make another type of clinometer from wood or metal. This model also has a plumb-line, but its reference scale gives you the slope in percent.

Making the pendulum clinometer, model 2

2. Cut a 51 x 51 cm square board from a piece of wood, or build
one from strips of wood or metal.

- If you cut the board from wood, use a piece which is heavy
  enough to prevent warping. Plywood or particle board 1.5 to
  2 cm thick will usually be good to use.
- If you build the board from wooden or metal strips, be sure
  the finished board is square. Carefully join wooden strips at
  the corners. You may need to brace them at the back with
  diagonal wooden strips. Securely weld metal strips together.

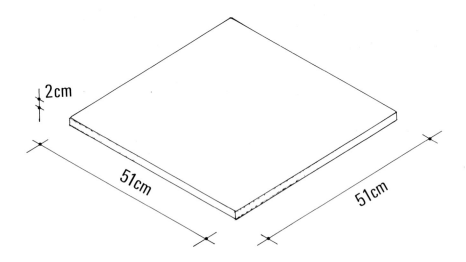

Reinforce the board if necessary

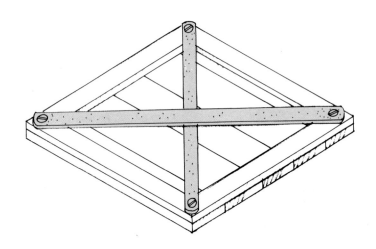

3. Provide a sighting line* along the upper edge of the square.

- If the board is made of wood, you can make a sighting line by driving a finishing nail (small-headed) into the top edge of the board at a point 2 cm in from each vertical edge. To make sure **the nails are at the same height**, place a block of wood 1.5 to 2 cm thick next to the first nail, and drive the nail in so that it is even with the top of this block. Then move the measuring block to the other end and use the same method for the second nail.
- If the board is made of metal, you can make a sighting line by glueing or welding two nails or metal points on the top edge of the board. Be certain that **the nails are at the same height**.

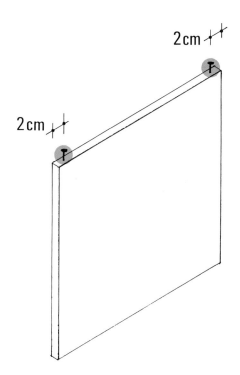

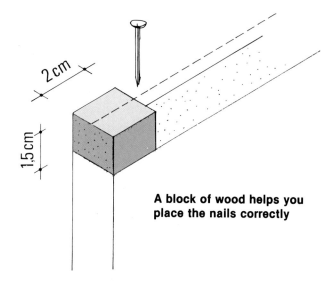

A block of wood helps you place the nails correctly

187

4. Provide a **centre-point from which to hang a plumb-line**. Put a mark on the board 1 cm down from the top edge and 1 cm in from the sighting point which is furthest from your eye. If the board is wooden, drive a nail into this mark; if it is metal, weld a small nail to the mark, or drill a hole through it.

5. Make a **plumb-line** about 65 cm long, using a piece of thin string and a weight. A plumb-bob (a small lead weight) will make the best weight for the plumb-line but, if you do not have one, you can use any object which has its weight evenly distributed from a single point. A heavy nut or washer, or a wooden disk with a hole in the centre, will work.

6. Attach the plumb-line to the hole or nail at the centre-point of the board.

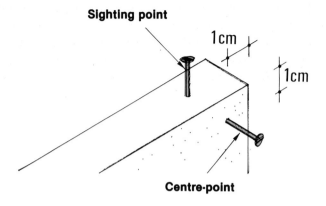

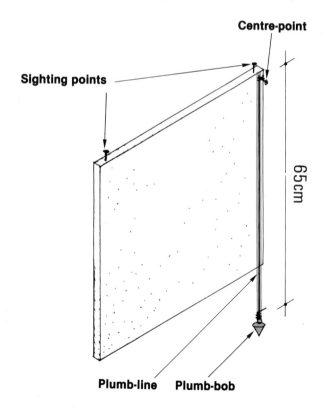

7. Loosely attach a **ruler graduated in centimetres** along the bottom edge of the board; use large clips, or tie the ruler on with string. Position the ruler so that its zero graduation is directly under the centre-point. Make sure that the distance between the centre-point of the plumb-line and the zero mark on the bottom edge of the ruler is 50 cm.

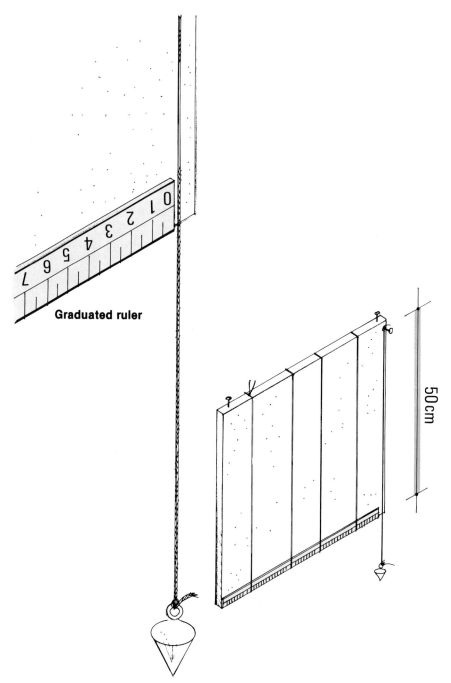

Graduated ruler

50 cm

Adjusting your clinometer

8. Aim the board at a mark which you have aligned at eye level (see Section 41). Standing straight and **looking along the board's upper edge**, align the two sighting points with this mark. Your **sighting line*** should now be horizontal and your plumb-line should be vertical.

9. Put your thumb on the plumb-line to hold it against the ruler at the bottom of the board, and check to see if **the line is at zero**. If it is not, adjust the position of the ruler so that the zero graduation and the plumb-line fall exactly in line.

10. Check to see that your clinometer is correctly aligned by sighting again. When it is, glue or nail **the ruler firmly in place**. Your clinometer is now ready to use.

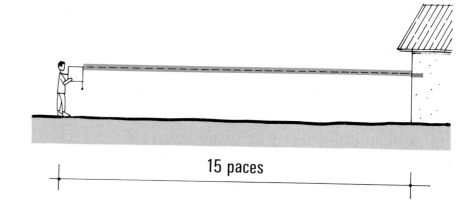

15 paces

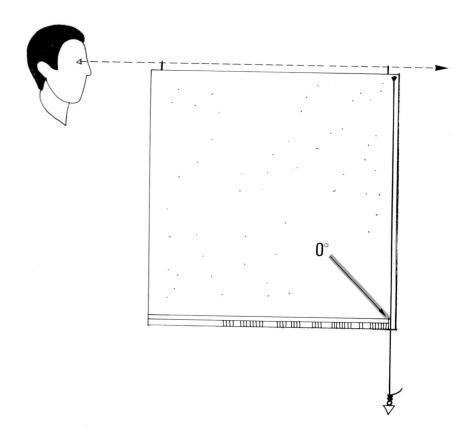

Using the clinometer to measure a slope

11. You can measure both uphill and downhill slopes with your clinometer in the following ways:

- To measure uphill slopes, the plumb-line should be at the edge of the board furthest from your eye when you are sighting;
- To measure downhill slopes, the plumb-line should be at the edge of the board nearest to your eye when you are sighting.

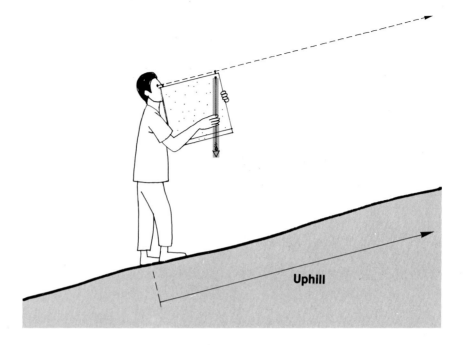

Uphill

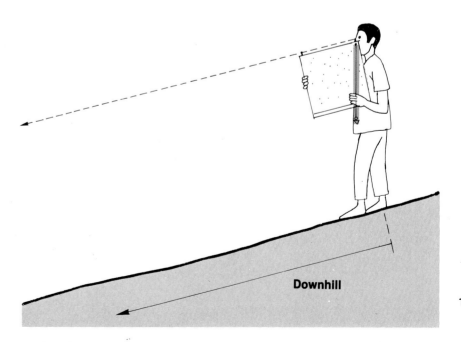

Downhill

191

12. Place a pole or a staff clearly **marked at eye level** (see Section 41, step 5) on a point you can easily see, usually 15 to 20 m away.

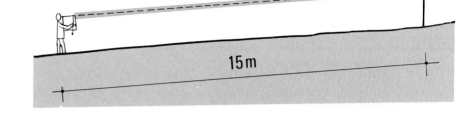

13. Aim the clinometer at this mark and, when the plumb-line has stopped swinging, press it with your finger to the ruler at the bottom. Be careful not to move the plumb-line from its vertical position. Then, read the graduation (in centimetres) at this point.

14. Since every centimetre on the ruler equals 2 percent of slope, calculate the slope as a percentage by **multiplying the number of centimetres** you read on the graduation **by 2**.

Example

If you read 2.5 cm on the ruler, the slope is found as:

> **2.5 cm × 2 = 5%**

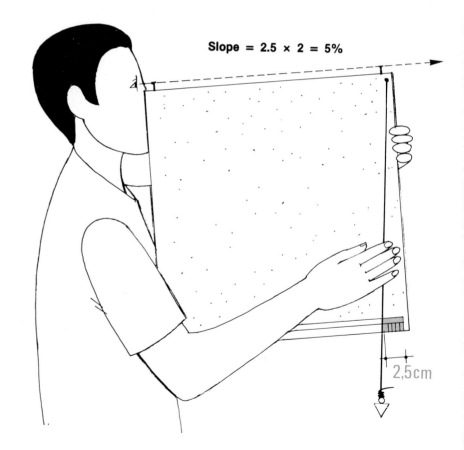

Slope = 2.5 × 2 = 5%

2,5 cm

43   How to measure with the home-made clinometer, model 3

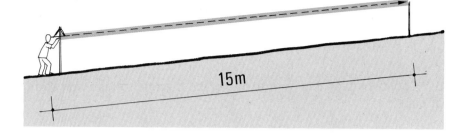

1.   The third model of clinometer is a little more complicated to make, but it is more accurate. It is also easier to use if you are measuring on ground that is soft enough for you to drive in the supporting staff.

Making the clinometer, model 3

2.   To make **the supporting staff**, get a straight stick or a piece of wood about 2 m long. Shape one of its ends into a point, so that you can easily drive it into the ground. About 25 cm from the pointed end, mark a line to show how deep you will drive the staff in.

3.   Get three pieces of wood exactly the same, 40 cm long, 4 to 5 cm wide, and about 1 cm thick. Secure them tightly together with nails or screws to form a **triangle with three equal sides**.

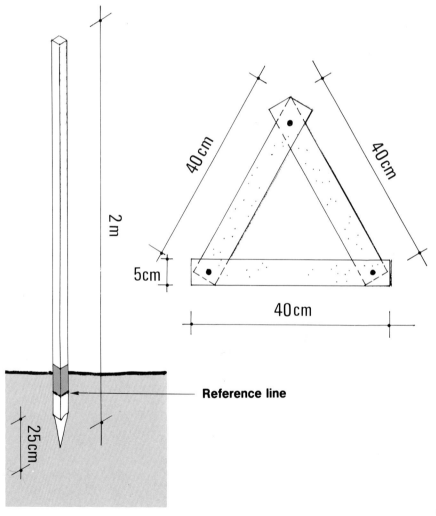

4. Prepare **a ruler graduated in millimetres**. Get a piece of wood about 25 cm long, 4 cm wide and 0.25 cm thick. Mark the centre with 0, then mark graduations from this centre-point up to 100 mm on either side.

5. Loosely attach this ruler to one of the triangle's sides with string or clips.

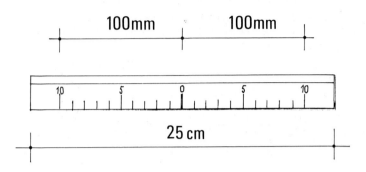

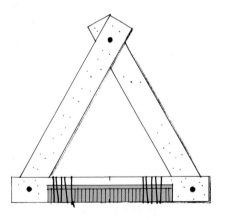

Tie the ruler to the triangle with string

6. On the same side of the triangle, make a **sighting device**. Drive two nails vertically into the side near each of its ends. Make sure the nails are at equal heights and on the line.

7. Drill a small hole exactly at the centre of the triangle's summit, opposite the zero point of the ruler.

8. Attach the triangle near the top of the supporting staff with a nail; make sure that the triangle remains free to swing around this axis.

9. Prepare a **plumb-line** about 40 cm long (see Section 42). Attach it to the nail at the centre of the triangle's summit.

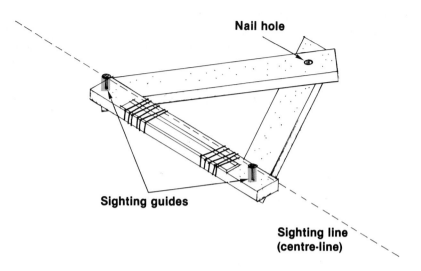

Nail hole

Sighting guides

Sighting line
(centre-line)

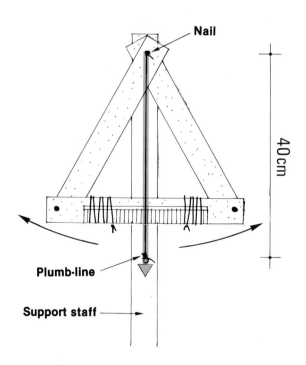

Nail

40 cm

Plumb-line

Support staff

Nail the triangle so it swings freely

195

Adjusting the clinometer

10. Drive the supporting staff **vertically** into horizontal ground until you reach the **reference*** level you marked above its pointed end.

11. Measure the vertical distance between the ground and **the sighting line*** of the clinometer exactly. This distance should be about 130 cm. Prepare a pole or staff that shows this height (see Section 41, step 5).

Note: the height of the sighting line for this clinometer may be different from your eye level.

12. About 15 paces away, make a mark on a wall set at the same height you just measured (see Section 41, step 5). Aim with the sighting line at this mark.

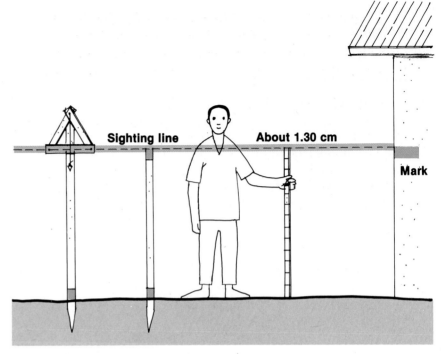

Sighting line   About 1.30 cm

Mark

Clinometer Sighting pole

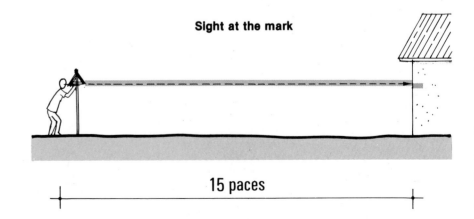

Sight at the mark

15 paces

13. Adjust the position of the ruler so that its **0-graduation** lines up exactly with the plumb-line. Check again for sighting-line accuracy and adjust the 0-graduation if you need to, then glue or nail the ruler firmly in position on the triangle. The clinometer is now ready to use.

14. Exactly **measure the distance** (in centimetres) between the point at which the plumb-line is attached and the point where the sighting line intersects the plumb-line. This distance should be about 32 cm, and is the **standard distance D** of your clinometer. Be sure to measure D precisely.

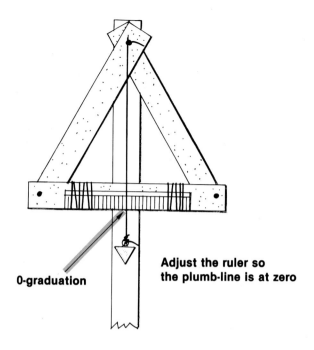

0-graduation

Adjust the ruler so the plumb-line is at zero

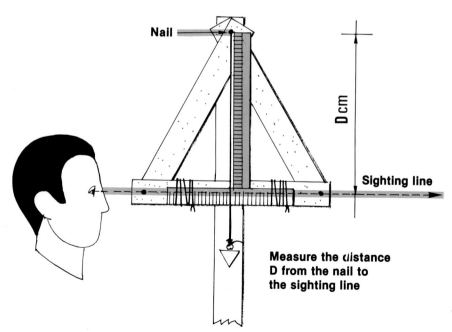

Nail

D cm

Sighting line

Measure the distance D from the nail to the sighting line

Using your clinometer to measure a slope

15.   You can measure either uphill or downhill slopes by reading the appropriate one of the two scales.

16.   Place a pole or staff clearly marked at the **sighting-line level** (see step 11) on a point B of the slope you are measuring, about 15-20 m away.

17.   At point A, drive your clinometer support **vertically** into the ground, down to the reference level. With the sighting line, aim at the mark on the pole or staff; to do this, slowly swing the triangle around the nail at its top until you sight the marked level.

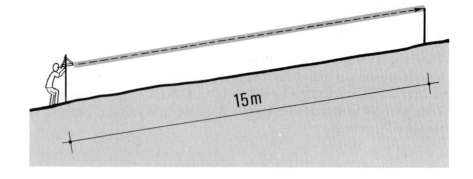

15m

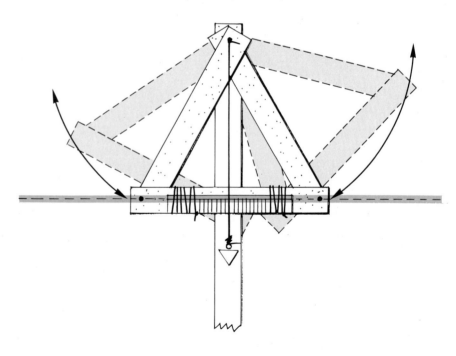

Swing the triangle around until you sight the top of the pole

18. When the sighting line is level with this mark, press the plumb-line with your finger against the ruler. Be careful not to move the plumb-line from its vertical position.

19. Read the **graduation N** (in millimetres) on the ruler at the point where the plumb-line intersects the sighting line.

20. If the **standard distance** of the clinometer (see step 14) is D (in centimetres), calculate the **ground slope S%** as:

$$S\% = (10 \times N) \div D$$

Example

If **D = 32 cm** and you read a graduation of **4.8 cm = 48 mm** on your clinometer, the slope is equal to:

$$(10 \times 48) \div 32 = 15\%$$

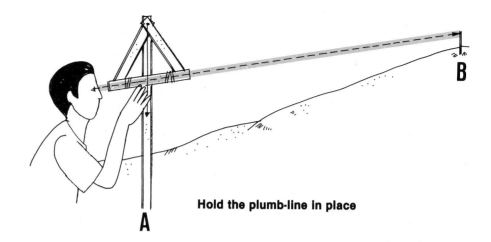

Hold the plumb-line in place

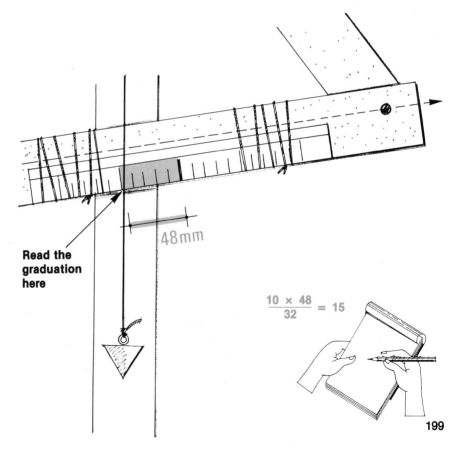

Read the graduation here

48 mm

$$\frac{10 \times 48}{32} = 15$$

199

1.   The fourth clinometer model is similar in principle to the preceding one, but it has several improvements: it is much smaller in size; it is easier to make; and it provides a direct reading of the slope, so that you do not need to make any calculations. The model 4 clinometer may also be used to measure vertical angles (see this Section, step 17).

Making the clinometer, model 4

2. Get a small piece of thin wooden board, about 14 x 21 cm. The best material would be plywood.

3. On this board, glue a sheet of squareruled millimetric paper so that its printed lines are parallel to the sides of the board.

4. Draw a line AB, parallel to the larger edge of the board and about 1.5 cm from it.

5. Find the **centre of line AB** and mark it C. From this point lay out **perpendicular CD**, which should measure 10 cm. You may adapt one of the methods from Section 36, or use the lines on the paper to guide you.

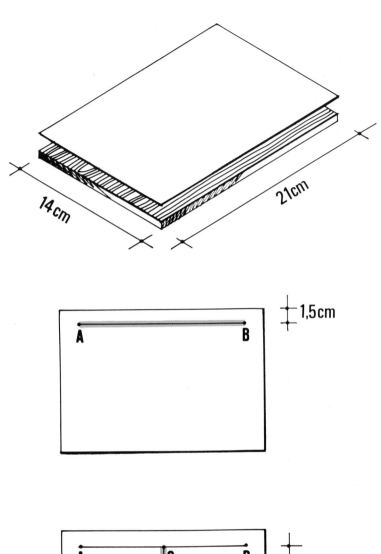

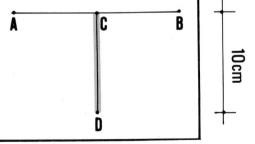

6.   Through point D, raise **perpendicular* EF**, which is **parallel*** to AB.

7.   Taking **point D as zero**, measure 10 cm to the left and 10 cm to the right of point D, along EF. Divide these two distances into **millimetres** and mark the main graduations. Once again, the lines on the paper will help you.

Note: instead of drawing the above lines yourself, you can use **Figure 4**. Either cut the figure directly along the dotted lines, or make a photocopy of it and cut it out. Glue this figure to the wooden board, with **line AB** parallel to the board's longer edge.

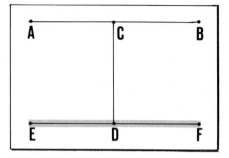

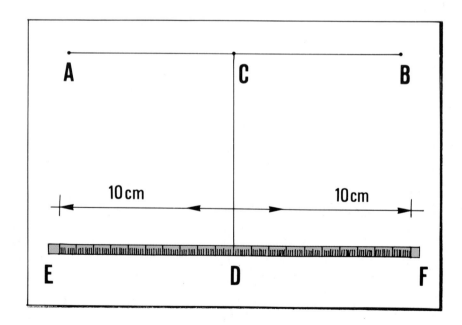

FIGURE 4

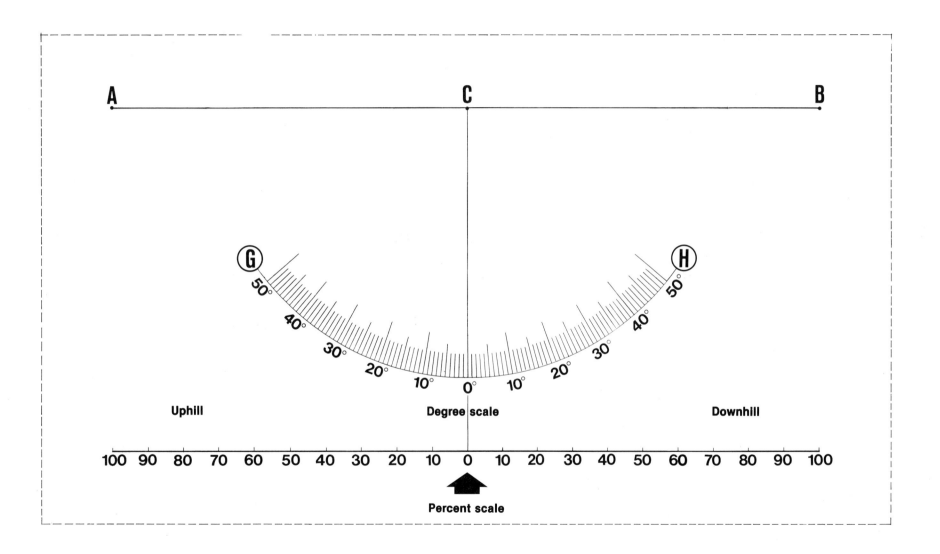

8. Make a **plumb-line** 17 cm long, using very thin thread (such as a nylon fishing line) and a small weight. Drive in a small nail exactly at point C on the board, and hang the plumb-line from it. Slightly below the nail, at K on line CD, drill a hole that a wood-screw will pass through.

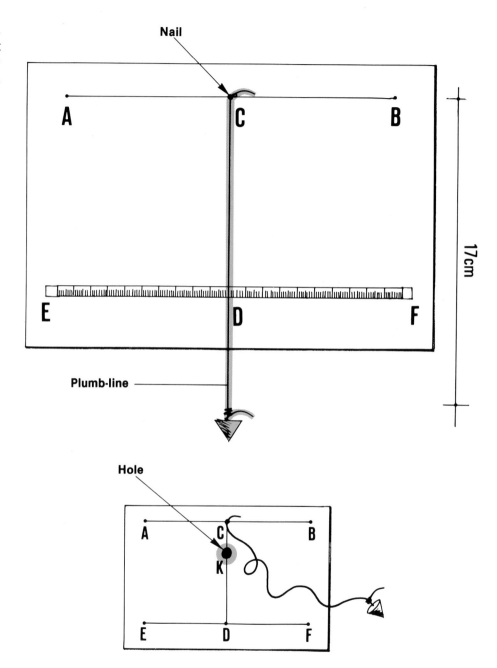

Nail

A   C   B

17 cm

E   D   F

Plumb-line

Hole

A   C   B

K

E   D   F

9. Make a **sighting line*** along line AB. To do this, you can drive thin nails in at points A and B. Or, get two metal strips (you can cut them from a tin) and cut small, v-shaped notches out of one end of each strip. Then, bend the other end so that the strips can be attached perpendicular to the board. Screw them to points A and B, making sure that the v-notches (your sighting guides) are directly over the two marked points A and B. Align these v-notches with line AB.

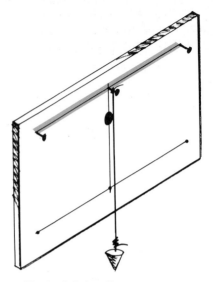

**Mark sighting line
AB with nails**

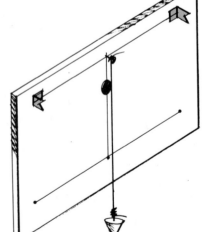

**...or with v-notch
sighting guides**

10.   Get a wooden staff 2 m long to use as the support, and make a point on the bottom end. Loosely attach the clinometer board near the top of this staff with a screw through the hole K you made on line CD in step 8. Tighten the screw so that the board can be turned around. Check that the head of the screw lies slightly **below** the surface of the board so it will not disturb the plumb-line.

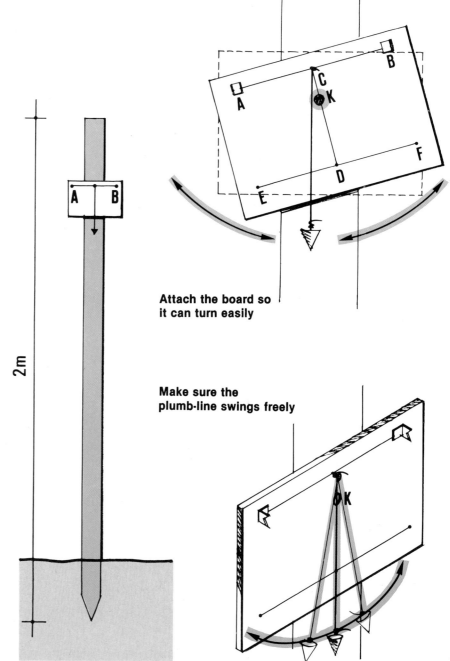

Attach the board so it can turn easily

Make sure the plumb-line swings freely

207

11. Clearly mark a **reference line*** about 25 cm above the pointed end of the supporting staff, showing the depth to which you need to drive it into the ground at each station. Measure the distance between this reference line and the sighting line AB.

12. Then prepare a pole or staff with a reference line and a sighting line at exactly the same height as line AB. This will be your **sighting pole**.

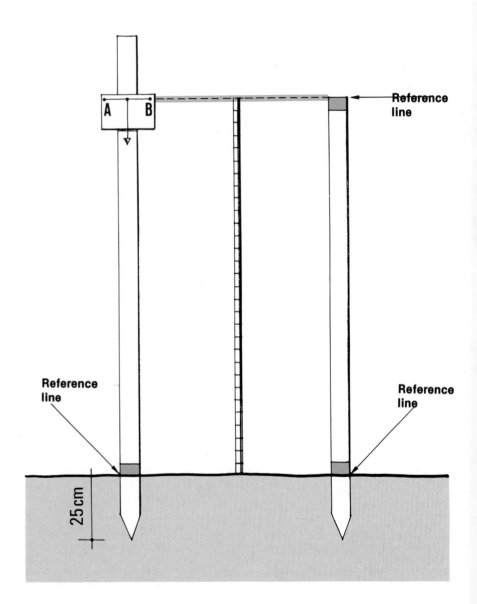

Using the clinometer for measuring a slope in percent

13.   You can measure either uphill or downhill slopes by reading the appropriate one of the two scales.

14.   Place the sighting pole you made in step 12 on point Y of the slope you are measuring, about 15-20 m away. Drive it in vertically up to the reference line.

15.   At point X, drive your clinometer support **vertically** into the ground up to the reference line. With the sighting line, aim at the mark on the sighting pole. Rotate the board around its screw until you sight the marked level.

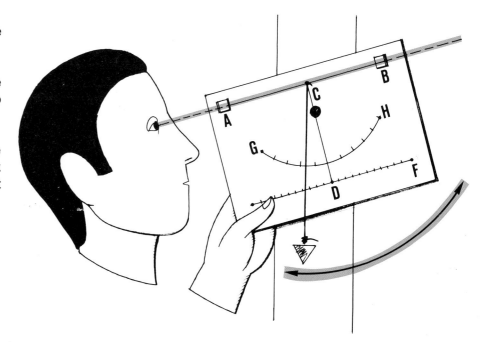

Turn the board until you sight the top of the pole

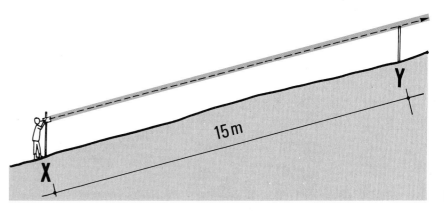

209

16. Where the plumb-line crosses line EF, read the graduation (in millimetres). This gives you **the slope in percent.**

Note: check carefully to see that the **plumb-line** hangs freely from its support. The board should rotate without disturbing the **vertical position** of the plumb-line.

Using the clinometer to measure a vertical angle in degrees

17. If you must measure a **vertical angle in degrees** instead of a slope, you may use the model 4 clinometer (as described above). The only difference in this case is that you **use the curved scale GH (in Figure 4)** rather than the bottom scale.

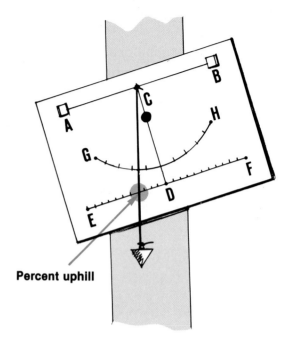

Percent uphill

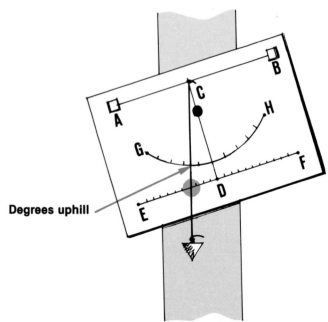

Degrees uphill

45   How to use the clisimeter

1.   The **clisimeter** is a simple instrument for measuring horizontal distances, as explained in Section 27. It can also be used to measure a slope or a vertical angle, but it can only give a **rough estimate** of these, accurate to within 10 percent.

The **lyra clisimeter** is a commonly used model. It is made up of a **sighting device**, an **attached ring**, and a **weight**, shaped like a pear, which keeps the clisimeter in a vertical position when hung from its ring. The instrument folds neatly into the weight for transport.

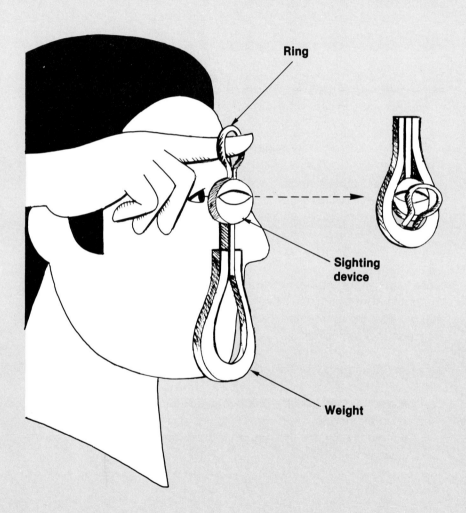

Ring

Sighting device

Weight

2.   When you look through the sighting device, you see three scales. As described before (see Section 27, step 3), **the central scale** is used to measure horizontal distances. The other two scales are used to measure vertical angles and slopes. You will use **the left scale**, which is graduated in **per thousand** (‰) or **tenths of percent** (%):

100 on the scale ‰ = 10%

or

5% = 50 on the scale ‰

Examples

15 per thousand equals   15 ÷ 10 =   1.5 percent
35 per thousand equals   35 ÷ 10 =   3.5 percent
150 per thousand equals 150 ÷ 10 = 15   percent
7 per thousand equals   7 ÷ 10 =   0.7 percent

Note: the right scale is graduated in **grades** (G), a unit of measurement which you have not used yet. A full circle is divided into 400 grades. Up to now we have been using **degrees**. There are 360 degrees in a full circle!

3.   The left scale is graduated from zero in two opposing directions:

- above zero are the positive **graduations** (+) for measuring uphill slopes;
- below zero are the **negative graduations** (—) for measuring downhill slopes.

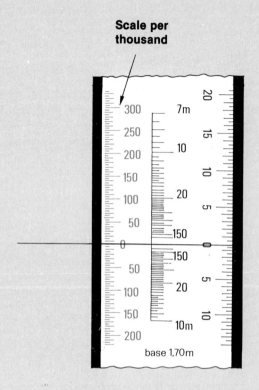

Scale per thousand

The scale inside the clisimeter — use the left scale to measure slopes

Using the clisimeter to measure a slope

You can use the clisimeter by yourself or with an assistant:

4. If you are working alone, you need a pointed stake clearly marked at two levels: the reference level above the pointed bottom, showing the depth to which you will drive the stake into the soil; and the **eye level**, which is the vertical measurement from the **reference level** to your eye level. It is best to have the eye level at the top of the stake. (This stake is like the one you learned to make in Section 41, step 5.)

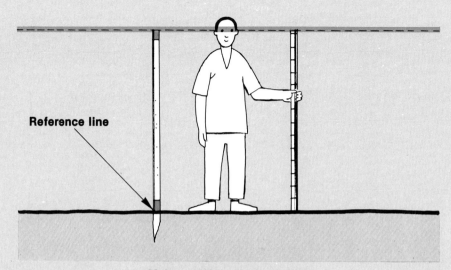

Reference line

Make a sighting pole marked at eye level

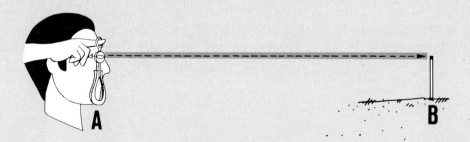

Sight at the marked pole

5. If you have an assistant, you can also use a simple rod marked at eye level, but it will be faster to use your assistant instead of this rod. To do this, determine the point on your assistant which is at the same level as your own eyes and sight at that point instead.

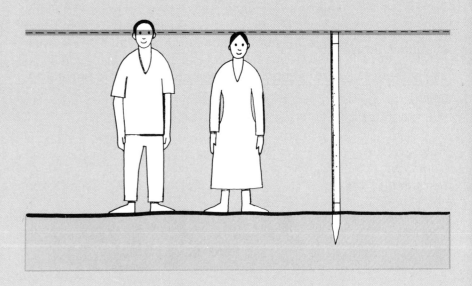

**Find your eye
level on your assistant**

**Sight at the eye
level you have chosen**

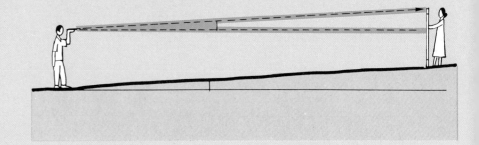

6. Place the **marked stake** at point B on the slope you need to measure, or send your assistant to point B, either with or without a marked rod.

7. Taking a position at point A about 10 to 15 m away, hang the clisimeter vertically from your left forefinger and bring the sighting device up to your left eye. Make sure to stand up straight so you do not change your eye level.

8. While looking at the marked level with your right eye, read the graduation on the left scale of the sighting device. This is the slope you are measuring, expressed in **per thousand.**

Note: to make reading the graduation easier, move your head slightly from right to left. The graduation will seem to extend out of the instrument into the landscape. Then, read the graduation corresponding to the marked level.

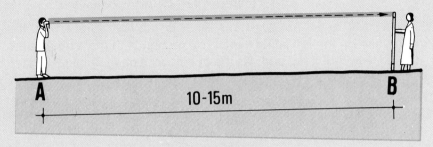

Sight at the point

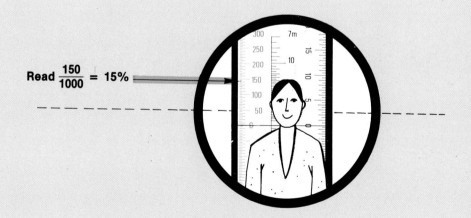

Read $\frac{150}{1000}$ = 15%

Read the left scale

Using the clisimeter to lay out a slope

9. You will need an assistant for this method. Sight with the graduation on the left scale (which corresponds to the slope) at the marked level (on a rod such as the one described in Section 41, step 5, for example) corresponding to the height of your eyes.

10. Ask your assistant to move the marked rod forward or backward until the eye level line is even with the clisimeter graduation.

11. When the rod is properly aligned, ask your assistant to mark the point on the ground with a stake. Move up to this stake and repeat the procedure.

Note: if you need greater accuracy, you can hang the clisimeter at a fixed height from a stick. If you do this, remember to adjust the marked level on the rod to this height.

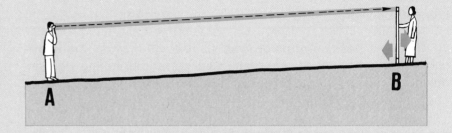

Move the sighting pole until you see it at the correct graduation

Read here

46   How to use the optical clinometer

1.   An **optical clinometer** is a precise pocket instrument for measuring vertical angles and estimating tree heights. It is commonly used by foresters. It can also be used to measure slopes quickly, with a method similar to that described for the clisimeter (see Section 45).

2.   When you look through the sighting device of the clinometer, you can see a cross-hair and two scales. The left scale is graduated in **degrees** and the right scale is graduated in **percent**. Both scales have a positive (+) section for measuring uphill slopes and a negative (—) section for measuring downhill slopes.

3.   Keeping both eyes open, sight with one eye through the optical clinometer, moving it until the cross-hair lines up with the **marked level** you wish to measure (such as a rod). With the clinometer lined up in this position read the graduation at the cross-hair.

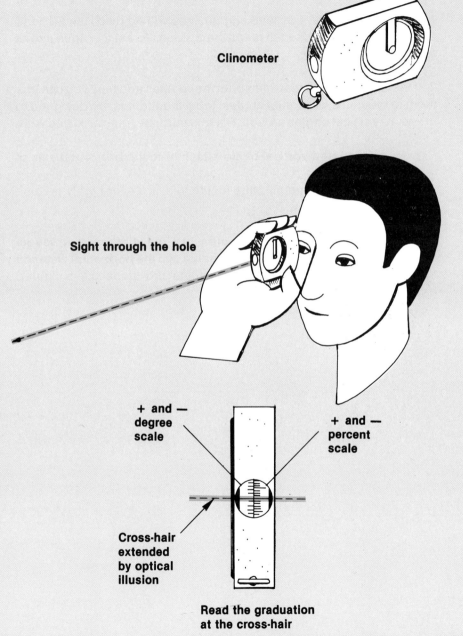

Clinometer

Sight through the hole

**+ and —
degree
scale**

**+ and —
percent
scale**

**Cross-hair
extended
by optical
illusion**

**Read the graduation
at the cross-hair**

47 How to use miscellaneous levelling devices

1. In Chapter 5, various levelling devices will be discussed. These devices can also be used to measure a slope. To set a graded line of slope, see Section 69.

2. In Section 35, you learned about **theodolites** and how you can use them to measure horizontal angles. Most theodolites are designed to measure **vertical angles** as well. For this purpose, they are fitted with:

- a **graduated vertical circle** attached to the horizontal axis of the telescope;
- an **extra graduated plate** inside this circle for highly precise measurements.

3. Levelling devices help you measure the **difference in levels** between two points. After you have measured the **horizontal distance** between these points, you can calculate the slope as explained earlier (see Section 40, step 8).

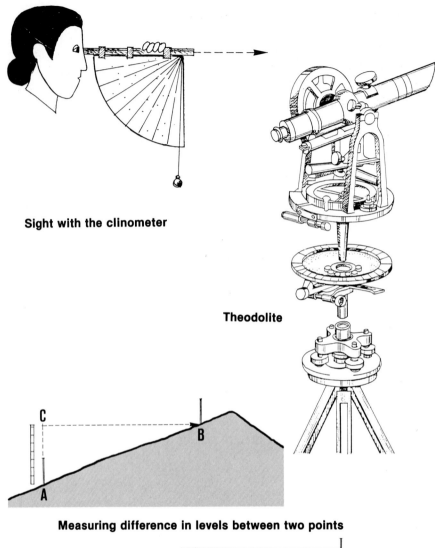

Sight with the clinometer

Theodolite

Measuring difference in levels between two points

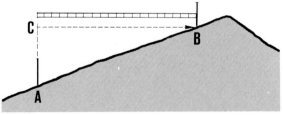

48 How to set out and check verticals

1. A vertical is a line with a 90° slope. You will often have to set out verticals, especially when you are building walls for a canal or building. You have already used vertical lines, to measure distances over sloping ground for example (see Section 26, step 19).

Setting out and checking verticals with a plumb-line

2. A **plumb-line** is a simple device which forms a **vertical line***. The idea of the plumb-line is based on the fact that any **heavy object will fall vertically**, making a 90° angle with the horizontal plane at ground level.

3. In a plumb-line, a fairly **heavy object**, the **plumb**, is attached to the end of a **thin line**. When the plumb hangs freely without moving, the line is vertical.

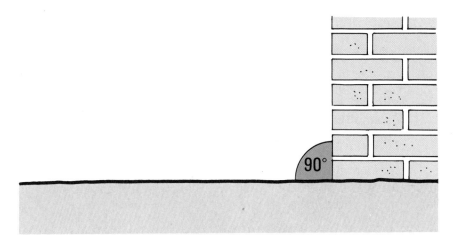

Most walls are vertical

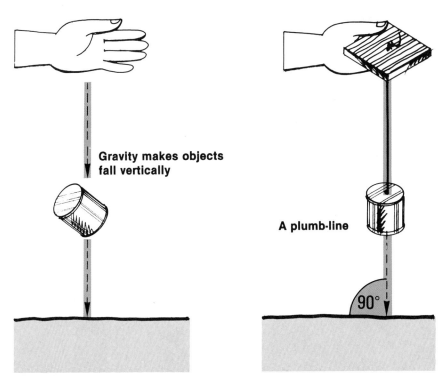

Gravity makes objects fall vertically

A plumb-line

Making your own plumb-line

4. You can make a **simple plumb-line** from:

- a **thin line** about 50 cm long, such as a piece of string, cotton thread, or nylon fishing line; and
- a **small but heavy object**, such as a stone, metal nut, or fishing lead.

5. You can make an **improved plumb-line** for measuring buildings in progress and other constructions. Start with a piece of wood or heavy metal about 10 cm square.

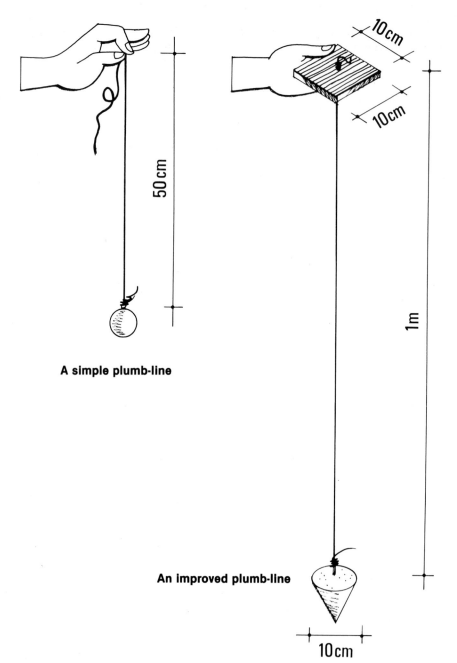

A simple plumb-line

An improved plumb-line

6. Find the exact centre of the square piece by drawing two diagonal lines on it. Drill a small hole through the point where they cross.

7. To make the plumb, get a heavy, solid block of wood (such as red acajou) or metal – the largest side of this block should be 10 cm across or less – if you can, shape the block into a cone.

8. If the block is wooden, drive a small nail into the exact centre of its top surface. If the block is metal, have a small hook welded to this point.

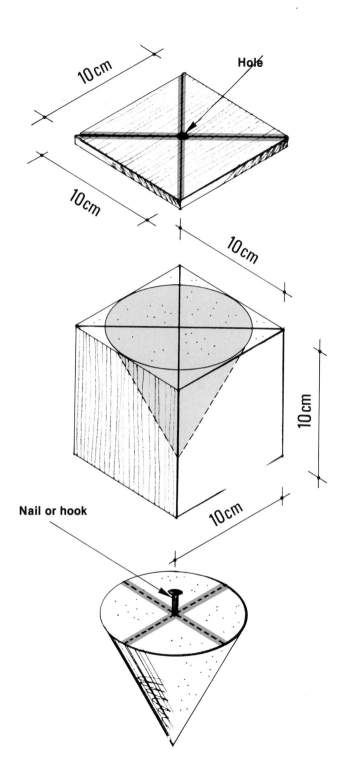

9. Attach the end of a thin line (nylon fishing line is strongest) about 1 m long to this nail or hook on the block and pass the other end through the central hole of the wooden or metal square piece. Fix the line on the other side of this hole either by tying it into a heavy knot or by tying a small piece of wood or metal (such as a nut) on to its end.

Note: you can change the dimensions of the plumb-line, depending on the materials you have. The line can be longer, if necessary.

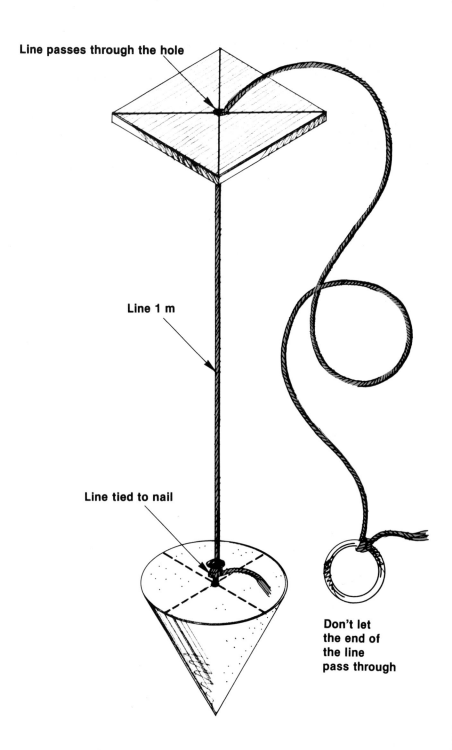

Line passes through the hole

Line 1 m

Line tied to nail

Don't let the end of the line pass through

Using a plumb-line to set out a vertical

10.   Remember that a free plumb-line will hang **vertically**.

11.   You can use a simple plumb-line to see if a wall is vertical. To do this, hold the top end of your plumb-line close to the wall and check to see if the distance between the wall and the top end of the line is equal to the distance between the wall and the centre of the weight at the bottom. This distance will be easier to check if the **weight is pointed** on the bottom.

Checking a vertical with a plumb-line

12. When using the improved plumb-line along a wall:

- if the diameter of the weight is equal to the diameter of the top square, place one of the sides of the square against the wall. Check to see that the side of the weight slightly touches the wall;
- if the diameter of the weight is smaller than the diameter of the top square, place one of the sides of the square against the wall. Check to see that the distance from the centre of the weight to the wall equals half the length of the square's side.

Note: if you need to make the plumb-line shorter to measure along walls of different heights, you can pull the line up through the centre-hole in the square at the top. Let it back down through the hole to measure higher walls.

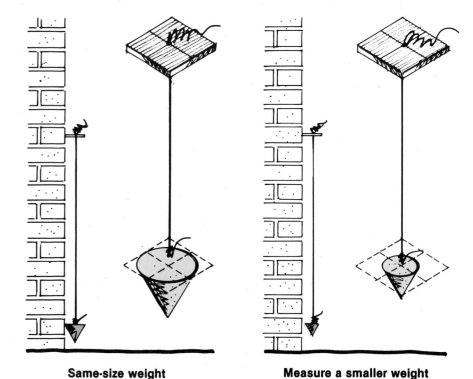

Same-size weight touches the wall

Measure a smaller weight

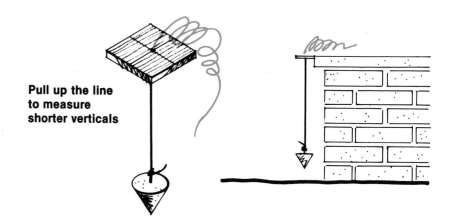

Pull up the line to measure shorter verticals

Checking small verticals with a mason's level

13. Some mason's levels (see Section 61) have an **additional bubble level for checking verticality**. You can use this level when you are building walls, for example. This method is particularly useful when the vertical you are checking is fairly small. Hold the mason's level vertically against the surface you need to check. If the surface is vertical, the bubble will be at the exact centre of the bubble level.

5 MEASURING HEIGHT DIFFERENCES – PART 1

Height differences in fish culture

1.   In fish culture, you must often measure the difference in height between two points. To construct a pond, you need to determine the heights of the dikes you will build, and the depths of the pond bottoms you will dig. To choose the routes of water-supply canals from the source to the ponds, you will also need height and depth measurements. And when you plan a reservoir, you will also need to make height measurements to determine where its shoreline will be (see Volume 4, **Water**, Section 42, page 102).

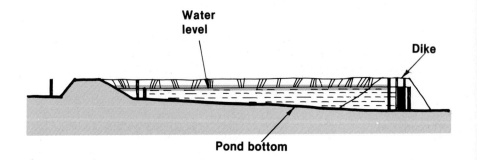

Types of problems will you be facing

1.   There are three types of problems in the measurement of height differences.

2.   You may have to measure **any differences in height** among a series of points on the ground, and compare them. From the results of this comparison, you can calculate the heights of given points so that you can make a map (see Chapter 9, in Book 2). This is called **surveying the levels** of the points, or **levelling** (see Sections 51-59).

3.   You may have to locate points which are at the **same height**. This is called laying out **contour lines**, or **contouring** (see Sections 62-68).

4.   You may have to locate points which have a **given difference in height**. In this case, you will be setting out **lines of slope** with a definite gradient (see Section 69).

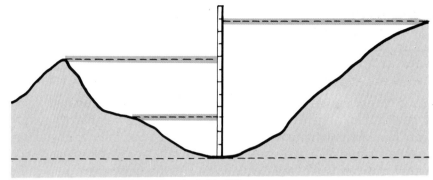

Calculating height differences

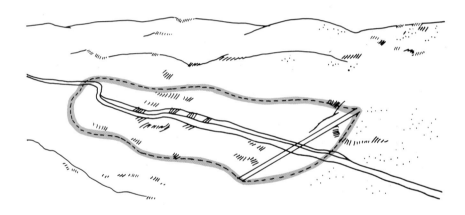

A contour line

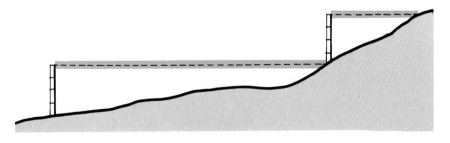

Setting out a slope

Measuring the height of ground points

5.   Differences in height between two points are usually measured with a device called **a level**. It is called a level because it gives **a true horizontal line**. The height of each point is then measured by its vertical distance above or below this horizontal line.

6.   This horizontal line can be formed in two ways, depending on the type of measuring device you are using to find the heights of points. If you use a **non-sighting level** (see Sections 51-53), the horizontal line will be formed by a straight-edge, a line level, or water levels. If you use a **sighting level** (see Sections 54-59), the horizontal line will be formed by prolonging a **line of sight***.

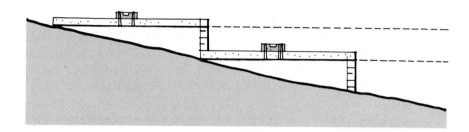

A straight-edge level

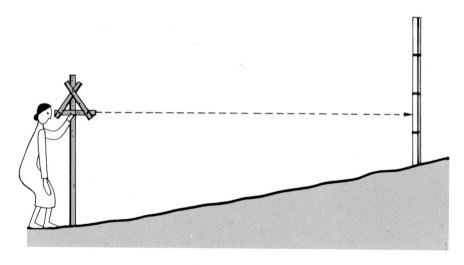

A sighting level

7. You will always use sighting levels together with a vertical graduated scale, **which measures the height** of the line of sight at each station.

8. A ruler with a vertical graduated scale is called a **levelling staff**. There are several models which you can buy, or you can make your own (see the following steps). Levelling staffs are usually 2 to 5 m long, foldable or telescopic, and made of plastic-coated wood or aluminium. **Self-reading levelling staffs** are usually graduated in metres, decimetres, and centimetres. These graduations are upside-down so that you can read them with a telescope. On a **target levelling staff**, there is a moveable "target" with a **reference line***, which can be positioned at a fixed height.

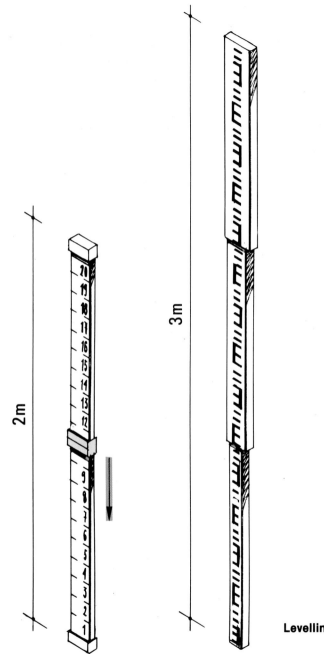

Levelling staffs

231

Making your own levelling staff

9.   To make **your own evelling staff**, get a straight length of wood, 2 to 3 m long and 5 to 10 cm wide. Clearly mark **graduations** along it every 10 cm. It is best if you paint the levelling staff white and mark the graduations on it in red. Make these graduations fairly large (about 1 cm thick) so you will be able to read them easily and accurately from a long distance.

10.   You can also make a levelling staff by gluing one or more **graduated measuring tapes** onto a straight piece of wood 2 to 3 m long. Glue the tapes on lengthwise, end to end. To read the small graduations accurately, you may have to decrease the distances over which you take the measurements, or rely on an assistant to make the reading.

11.   Other models of levelling staff are described in **Section 65**. These are used for contouring and for setting graded lines of slope.

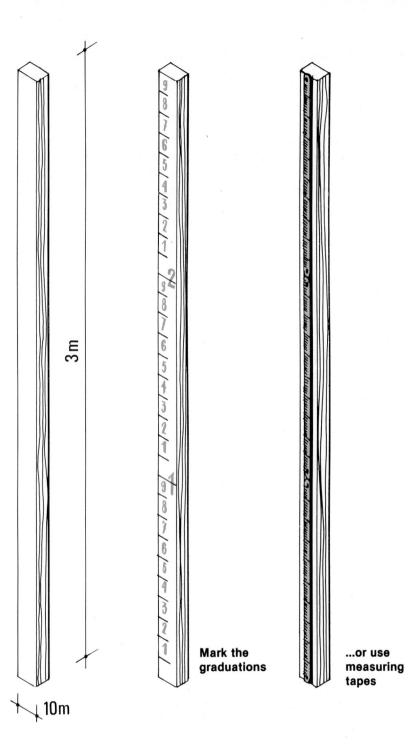

Mark the graduations

...or use measuring tapes

Choosing the best method for measuring height differences

12. There are a number of good ways to measure height differences. The method of measurement you use will depend on several factors. Each method is discussed in the following sections. **Table 7** will also help you to compare the methods and select the one best suited to your needs. Additional information on how to plan your levelling survey, how to record the measurements and how to calculate the results will be given in Chapters 7 and 8 in Book 2.

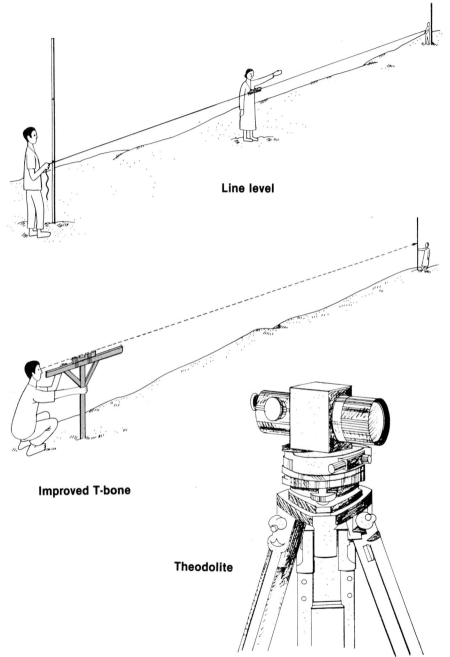

Line level

Improved T-bone

Theodolite

TABLE 7
Height difference measurement methods [1]

NON-SIGHTING LEVELS

Section [2]	Method [3]	Distance, m	Accuracy	Remarks	People, equipment
51 *	*Straight-edge level*	2.5 to 3	Medium to high	Easy to transport Quick to operate	1 person, mason's level 1 measuring scale
52 *	*Line level*	20	Medium	Very easy to transport Quick to operate For rough ground	3 people, mason's level 2 measuring scales
53 *	*Flexible tube water level*	10 to 15	High	Awkward to transport Very quick to operate For clear rough ground Avoid water loss	2 people, 2 measuring scales

SIGHTING LEVELS

Section [2]	Method [3]	Distance, m	Accuracy	Remarks	People, equipment
54 *	*T-bone level*	10	Low to medium	Rough measurement Useful for canals, pipelines	2 people, 1 levelling staff
55 *	*Improved T-bone level*	15 to 20	Medium	Especially good for dikes and future water level	2 people, mason's level, 1 levelling staff
56 **	*Bamboo level*	15 to 20	Low to medium	Greatly affected by wind	2 people, 1 levelling staff
57 *	Hand level	10 to 15	Low	Rough, quick method, best when rested on a pole	2 people, 1 hand level, 1 levelling staff
58 ***	Surveyor's level and theodolite	over 100	Very high	Expensive, delicate Automatic levelling with stadia hairs	2 people, expensive level, tripod, 1 special levelling staff

[1] See also Table 8 for contouring and slope setting

[2] * Simple   ** more difficult   *** most difficult

[3] *In italics*, equipment you can make yourself

Calculating height differences from slopes

13. If you know **the average slope** between two points (see Chapter 4), you can easily calculate the height difference between them. First measure **the horizontal distance D** in metres between points **A** and **B** (see Chapter 2). To calculate the height difference **H** (in metres), multiply **D** by the slope **S** expressed in hundredths:

$$H = D \times 0.0S$$

Example

- You measure **D** = 20 m and **S** = 5% = 0.05.
- **H** = 20 m × 0.05 = 1 m.

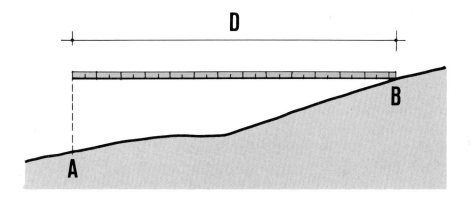

Measure the horizontal distance

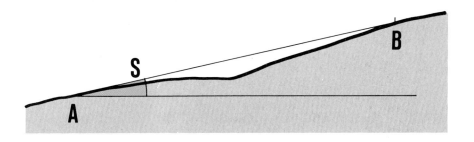

...and the slope

...to calculate the height

$$H = D \times 0.0 S$$

235

Calculating height differences from vertical angles

14.   If you have measured vertical angle ABC in degrees, you can calculate the height difference AC from:

- either the **ground distance BC** along the slope

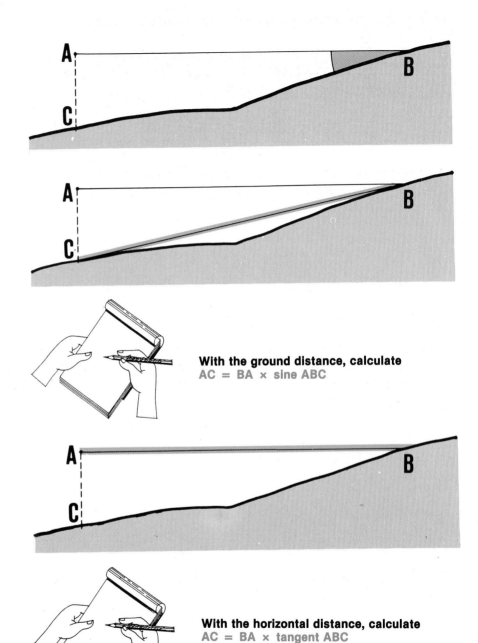

$$AC = BC \times \sin ABC$$

obtaining **sine ABC** from **Table 14;** or

- the **horizontal distance BA**

$$AC = BA \times \tan ABC$$

With the ground distance, calculate
AC = BA × sine ABC

obtaining **tangent ABC** from **Table 3.**

Example

- Vertical angle ABC = 7 degrees. You need to calculate AC.
- If you have measured BC = 47 m: from Table 14, sin $\overset{\circ}{7}$ = 0.12187
  AC = 47 m × 0.12187 = 5.72789 m = 5.73 m;
- If you have measured BA = 46.7 m: from Table 3, tan $\overset{\circ}{7}$ = 0.1228
  AC = 46.7 m × 0.1228 = 5.73476 = 5.73 m.

With the horizontal distance, calculate
AC = BA × tangent ABC

Using height differences to calculate horizontal distances

15.   You learned earlier that, **on sloping ground**, you need to correct distance measurements taken along the ground in order to find the horizontal distances (see Section 26).

16.   You also learned one way of correcting distance measurements, using **slope** (see Section 40, steps 10-13).

17.   Another way of correcting distance measurements is to use measurements of **height** differences in the following formula:

$$\text{Horizontal distance} = \quad G^2 - H^2$$

where **G** = AB is the distance measurement (in metres) along the sloping ground and **H** = AC is the height difference (in metres) between the two points.

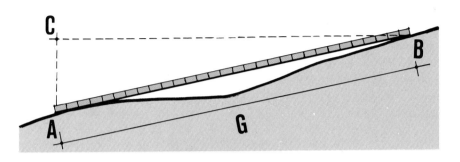

Ground distance

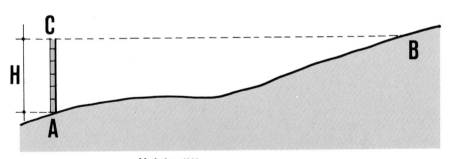

Height difference

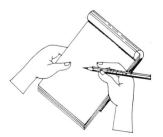

$$CB = \sqrt{G^2 - H^2}$$

Example

- You have measured AB = 45 m along the sloping ground;
- The height difference AC from point A to point B equals 9 m;
- The horizontal distance

$$CB = \sqrt{(45\ m)^2 - (9\ m)^2} =$$

$$= \sqrt{2025\ m - 81\ m} = \sqrt{1944\ m} = 44.1\ m$$

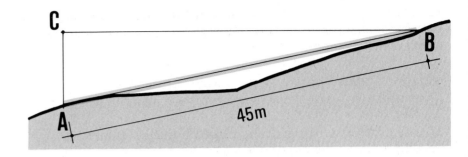

Ground distance = 45 m

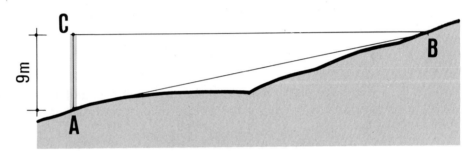

Height difference = 9 m

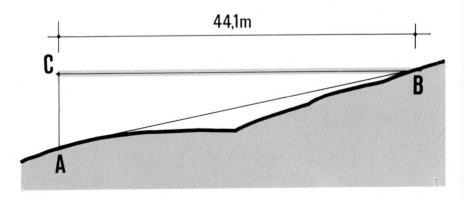

$$\sqrt{45^2 - 9^2} = 44.1$$

51 How to use the straight-edge level

1. You can make a simple device for measuring height differences over a small distance out of a mason's level (see Section 61) and a wooden straight-edge.

Making a straight edge

2. Get a piece of wood that is heavy enough to resist **warping**, and 2.5 to 3 m long. This piece of wood must be cut very carefully so that the edges are straight and squared-off.

3. When you have cut the straight-edge, hold one end to your eye and sight along the top and bottom edges to find which is the straightest.

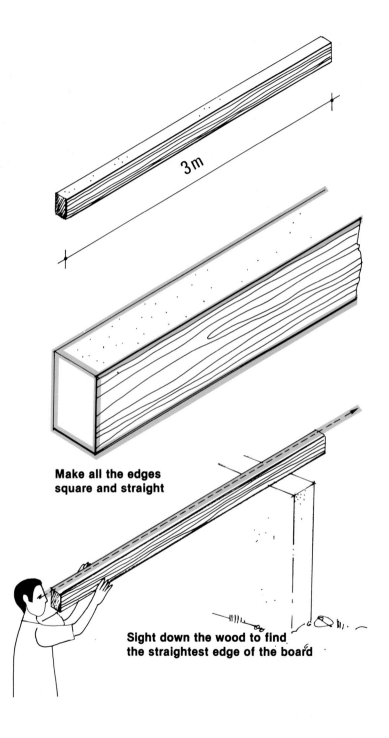

Make all the edges square and straight

Sight down the wood to find the straightest edge of the board

239

Making a straight-edge level

4.   Using light string, lash a mason's level securely to the mid-point of the straightest edge of your straight-edge. Make sure that the mason's level is parallel to this edge.

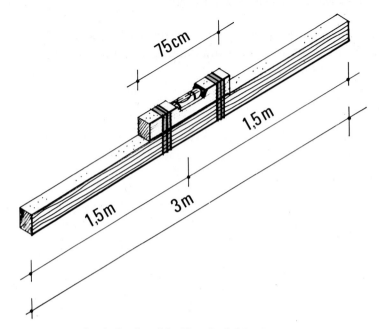

Lash the level to the straightest edge of the board

Measuring height differences using the straight-edge level

5.   Set one end of the straight-edge level on the ground at the highest point **A**, and move the other end up or down until the edge is horizontal, using **the spirit level** as a guide.

6.   Measure the vertical distance from point **B** on the ground to the bottom of the straight-edge level, using a graduated ruler, for example.

Note: if the distance AB between the two points is greater than the length AC of the straight-edge level, you will have to measure intermediate points C, D, E ... B and add up all the heights to get the total height.

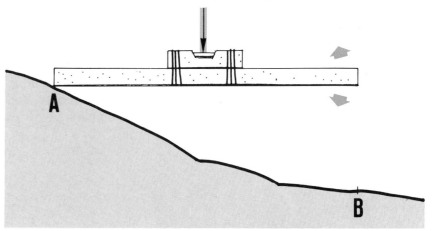

When the level is horizontal

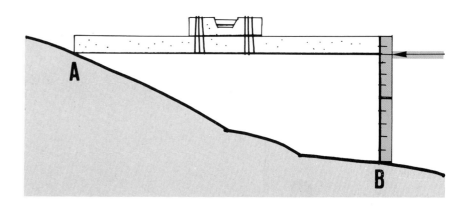

...measure the height difference

Measure big differences in stages

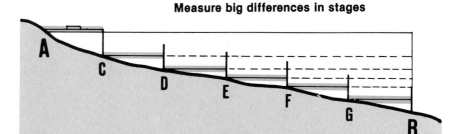

241

52 How to use the line level

A mason's level and a cord form the basis of the line level. It is a simple device which you can use over a relatively long distance (up to about 20 m). You will need to work in a team of three people. You will also need two levelling staffs and some marking pins.

Making your own line level

1. Get a cheap mason's level with a wooden case. Screw a strong screw-eye into each end-face, on the centre-line and close to the top.

2. Get two 10-m-long cords and tie one of them to each of the screw-eyes.

3. Wrap the loose ends of the cords with string to reinforce them.

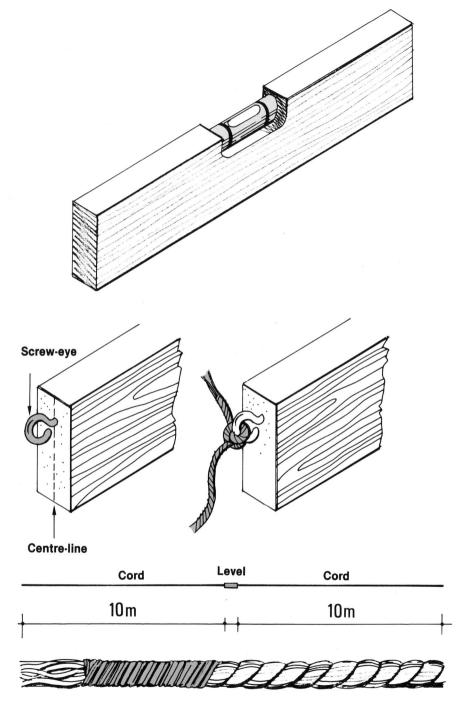

Screw-eye

Centre-line

	Cord	Level	Cord	
	10m		10m	

Wrap the ends of the cords for strength

Using the line level for levelling

4.   The **rear person** places a levelling staff on the starting point A of the line you are level-surveying. The end of one cord is put against the 1 m graduation of the staff.

5.   The **front person** then takes a levelling staff, a marking pin and the end of the other cord, and walks away from the **rear person**, following the direction of the line being surveyed, and stopping when the cord is well stretched.

6.   The **front person** places the second levelling staff vertically in the ground making sure that it is on the line being levelled. The end of the cord is pulled until the entire line level is as nearly **horizontal** as possible. This point is then marked with the marking pin.

7.   The **centre person** stands between the rear and front people and looks at the mason's level; then signals the front person to move the end of the cord up or down the levelling staff, **until the spirit level indicates that the line is horizontal**. If necessary, the rear person also moves that end of the cord up or down, to prevent the mason's level from touching the ground.

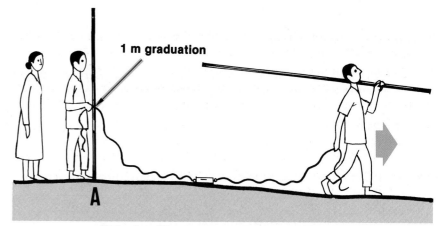

Put a levelling staff at starting point A

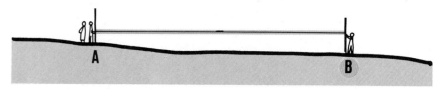

When the cord is stretched tight, mark point B

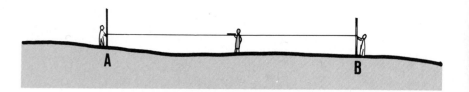

The centre person looks at the level

...and guides the front person until it is horizontal

8.   The rear person **reads the height of that end of the cord on the levelling staff**. The front person does the same. Be sure to double-check all measurements. Write down the measurements carefully, putting the rear measurements in one column and the front measurements in another column so you won't get them confused. (See the chart in step 10.)

9.   Then the front person removes the levelling staff from the ground and replaces it with a marking pin. The team progresses forward along the line, repeating the same procedure. The rear person should stop each time at the marking pin that the front person has placed for the previous measurement.

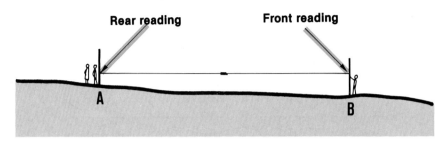

Read the heights on the levelling staffs

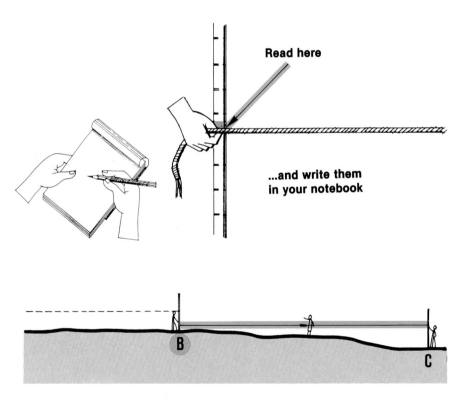

Read here

...and write them in your notebook

Begin the next measurement from point B

10. To calculate the height differences for the entire line, first find the height difference for each station by subtraction. Add up all the differences to find the total height difference (see below and Section 81 in Book 2).

Finding height differences with the line level

Station	Rear	Front	Difference
1	100 cm	96 cm	4 cm
2	100 cm	89 cm	11 cm
3	100 cm	92 cm	8 cm
1 to 3		**TOTAL:**	**23 cm**

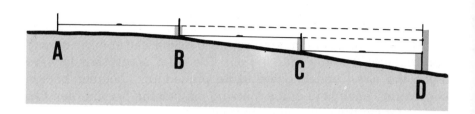

Add the measurements to find the total difference

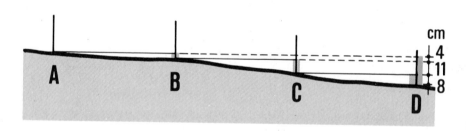

4 cm + 11 cm + 8 cm = 23 cm

246

53 How to use the flexible-tube water level

You can make a simple device for measuring level distances using a piece of transparent water hose 10 to 15 m long and two levelling staffs.

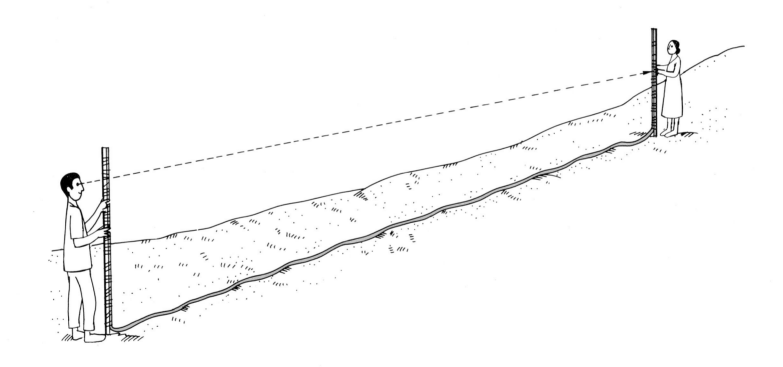

Making your own water level

1. If you do not have two levelling staffs, get two straight pieces of wood, 4 x 2 cm wide and 2 m long. Mark off a measuring scale in centimetres on each of them, or get two measuring tapes and glue them lengthwise to the pieces of wood.

2. When you mark the centimetre scales on the pieces of wood, place them side by side and align their tops and bottoms so that you can be sure both scales will be at exactly the same level. If you begin marking the scales 10 cm from the bottom of the wood, you can easily see where they begin, even if you are measuring in tall vegetation. Make sure that the bottom of each staff is flat or has a **reference line***.

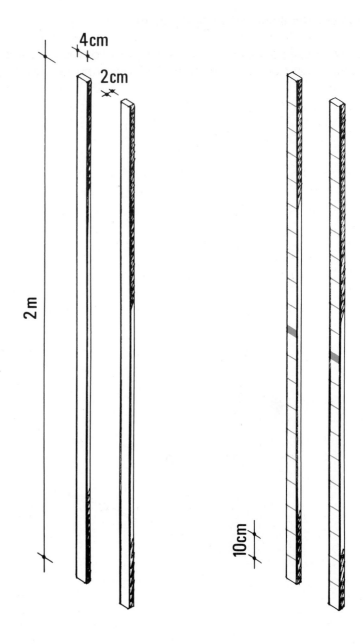

4 cm

2 cm

2 m

10 cm

3. Lay the two levelling staffs side by side in front of you, with their scales facing you. With strong string, lash the plastic hose along the length of the inside edges of the measuring scales. Make sure that the very ends of the hose are even with the tops of the staffs. The middle part of the hose will be loose between the two poles. When you fasten the hose to the poles, tie the string around the hose tightly enough to be secure, but take care not to pinch the hose. Make sure that the very ends of the hose are lashed to the scales.

4. On one point on the ground, place the two measuring scales side by side, with their scales aligned, in a vertical position. Slowly fill the hose with water, taking care to get rid of any air bubbles, until the level is about 1 m high in each of the upright sections of hose (the stand pipes) when they are held together.

5. Plug each end of the hose with a cork or another kind of stopper to avoid losing water when you carry the level. If you do lose any water, align the scales as you did before and refill with water to about 1 m.

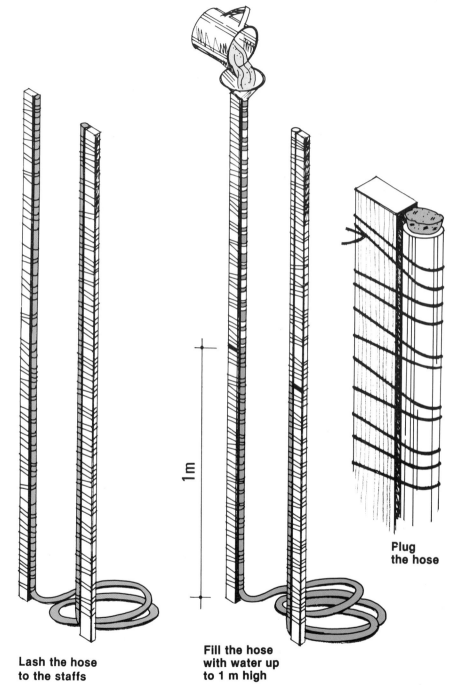

1m

Lash the hose to the staffs

Fill the hose with water up to 1 m high

Plug the hose

Using the water level for levelling

6. To use the water level, you will need to work in a team of two people. The rear person stands at the starting point A of the line, and places one of the measuring scales in a vertical position on the ground.

7. The front person, carrying the other measuring scale and a marking pin, walks ahead along the line in the direction of the point where you want to find the difference in levels. When the end of the hose is reached the measuring scale is placed in a vertical position on the ground. Make sure that the levelling staff is directly on the line.

8. When the measuring scales are in position, both people **remove the plugs** in their ends of the hose. This is to ensure that the water in the hose will fall at the correct level.

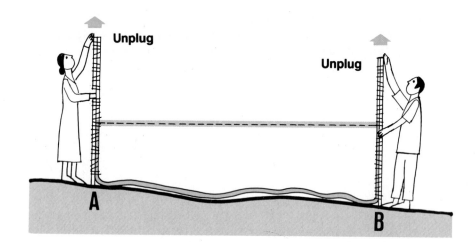

9. Read the measuring scales with your eyes level with the surface of the water in the hose.

10. Replace the plugs in the ends of the hose.

11. Note down the measurements in a special table which will help you to calculate the height differences accurately (see Section 52, step 10). The front person marks the point where he or she is standing with a marking pin.

12. Progress forward, repeating the same procedure along the line. Each time you finish a section, the rear person should take a position at the marking pin left by the front person.

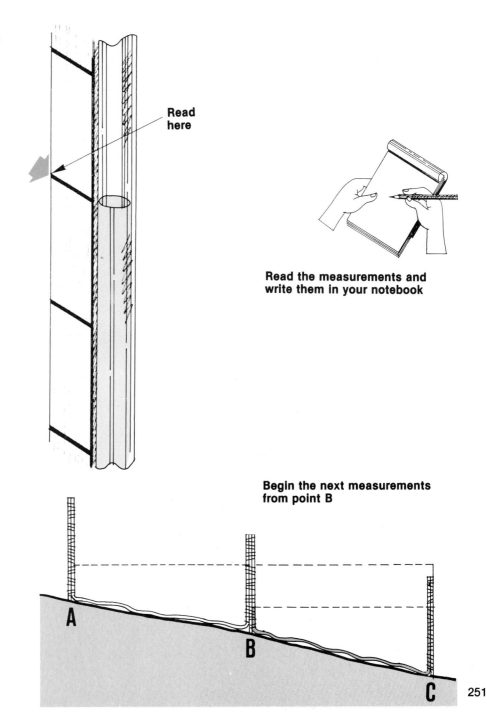

Read here

Read the measurements and write them in your notebook

Begin the next measurements from point B

A

B

C

251

54   How to use the T-bone level

The T-bone level is a very simple level which is particularly useful for
setting out canals or pipeline centre-lines. You use it together with a
levelling staff, held by an assistant.

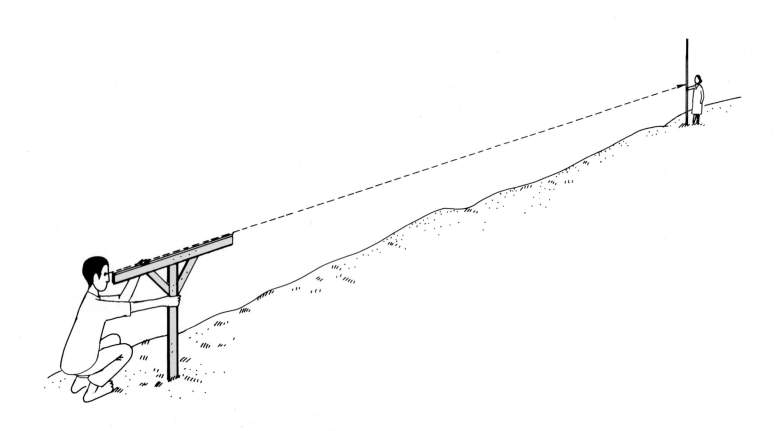

Making a T-bone level

1. Get two 5 × 2.5 cm pieces of wood, each exactly 1 m long.

2. Along the 2.5 cm face of one of these pieces, draw the centre-line. Make a shallow groove along this line with a saw.

3. Lay the other piece of wood lengthwise on the ground, and centre the grooved piece, grooved side up, perpendicularly across it in a "T" shape. Make sure that their top sides are even and that they form an exact 90° angle. Nail the grooved piece in position to the other piece and add two support struts to hold it in place. The total height of the device should be 1 m.

Note: to improve the accuracy of the level, you can make the horizontal top piece 1.5 m long rather than 1 m.

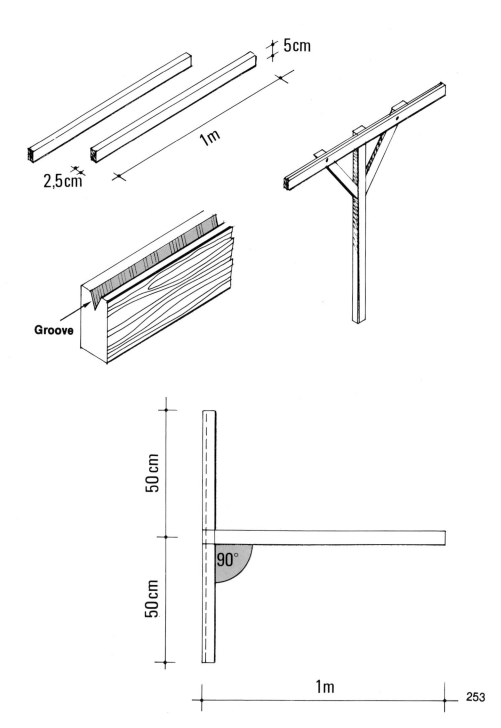

Groove

50 cm

50 cm

90°

1m

253

Using a T-bone level for levelling

4. At the starting point A of the line you are level-surveying, stand with the T-bone level. Hold the base of the level **firmly on the ground,** avoiding stones or other objects which might cause it to wobble. Be sure the support is held **vertically**.

5. Have your assistant hold the levelling staff in a vertical position at the next point B of the line, about 10 m away.

6. Sight **along the edge of the groove**, as though you were sighting a gun, toward the graduation of the levelling staff.

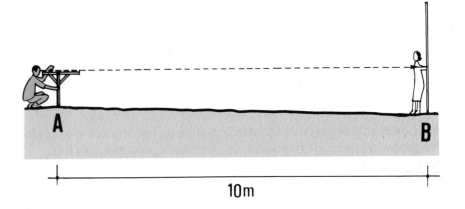

10m

Hold the level firmly in place

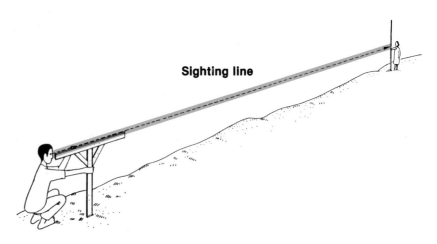

Sighting line

Sight along the groove at the levelling staff

254

7. On the levelling staff, read the height corresponding to the T-bone level sighting line and note it down. Your assistant can help you by slowly moving a brightly coloured marker, such as a pencil or a pen, up and down along the levelling staff until you signal that it is level with the T-bone top edge. He or she then reads the height to you.

8. Note these readings down in a table and calculate the height differences (see Section 52, step 10).

Note: since your T-bone level is exactly 1 m high, all you have to do to obtain the height difference between two points is to subtract 1 m from the reading on the levelling staff.

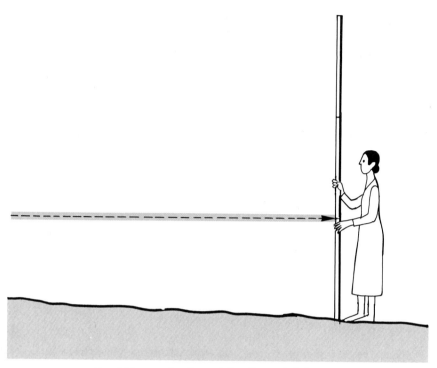

Read the graduation at the line of sight

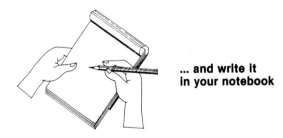

... and write it in your notebook

55   How to use the improved T-bone level

To make an improved T-bone level you can add a **mason's level** to the original device to help to make its sighting line horizontal. It can be used over longer distances, particularly to set out the top levels of pond dikes and to determine the water line of future reservoirs (see Vol.4, **Water**, Section 42).

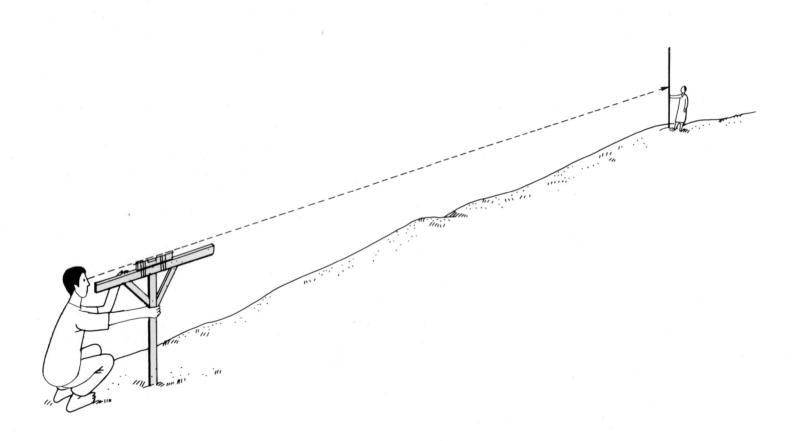

Making your improved T-bone level

1. Get a mason's level with a wooden case and attach two **metal sight pieces** to its ends. To make them, cut two strips from a flattened tin. They should be the same width as the narrowest part of the mason's level, and about 2 cm longer than its height. Cut a V-notch 1 cm deep in one end of each strip. Nail the strips to the ends of the mason's level with the notches sticking up to create a **line of sight** along the top of the mason's level.

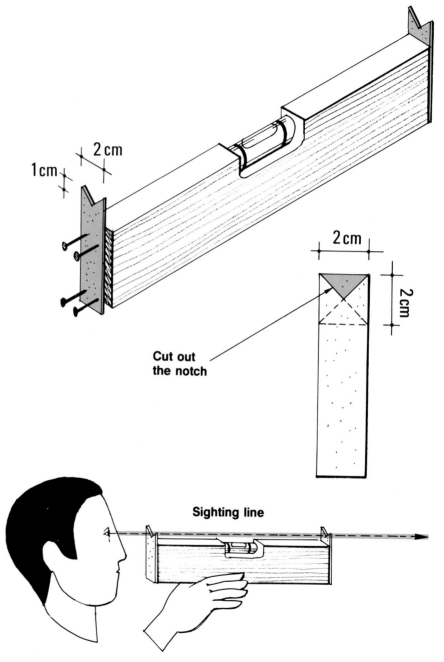

2 cm

1 cm

2 cm

2 cm

Cut out
the notch

Sighting line

2. Get two pieces of wood, each 5 x 2.5 cm thick and about 1 m long. Assemble them with wooden support struts so that:

- the top piece forms a 90° angle with the support piece and is centred over it in a T-shape; and
- the **widest face** of the top piece is horizontal to provide a flat surface.

3. Centre your modified mason's level on the top piece and attach it. Then measure exactly 1 m from the sighting line at the top of the mason's level down the support piece. Clearly mark this **reference line*** with paint or with a narrow piece of wood nailed across the support. Below this mark, shape the support into a point.

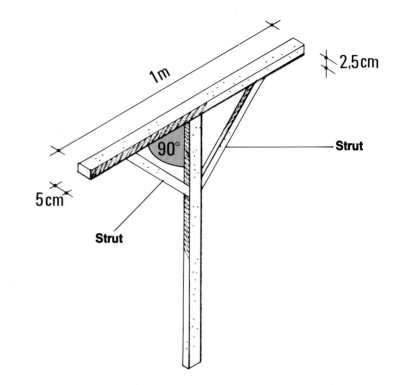

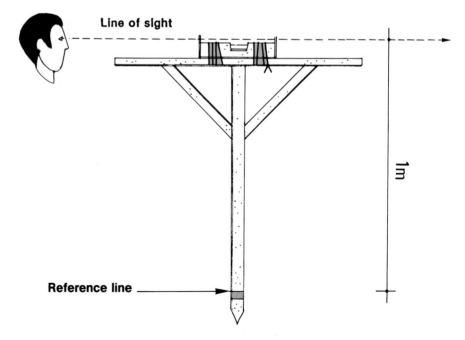

Using the improved T-bone level for levelling

4. You use the improved T-bone level the same way as the simple T-bone level (see Section 54), except that:

- first you drive it into the ground, down to the reference line;
- then you adjust the top board with the mason's level to make it horizontal;
- finally, you set the sighting line with the metal sights attached to the mason's level.

Note: the sighting line will be **exactly 1 m above** point A where the improved T-bone level is positioned. Knowing this, you can easily determine the other points B, C, ... G of the site that are 1 m higher than your levelling station A by standing on the same point and levelling around in a circle.

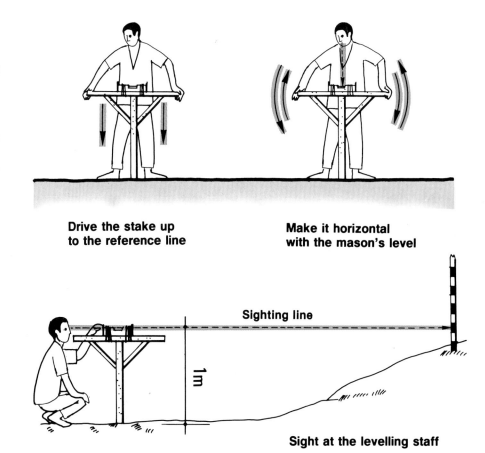

Drive the stake up to the reference line

Make it horizontal with the mason's level

Sighting line

1m

Sight at the levelling staff

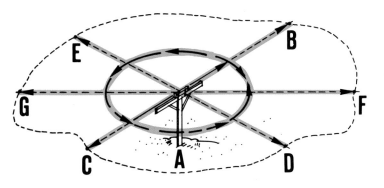

Rotate the level to find all the points 1 m higher than point A

56 How to use the bamboo sighting level

You can make a simple device for level surveying from a small bamboo tube and several pieces of wood. It should be used with a levelling staff. It is **very sensitive to wind and breezes**. When you use it, make sure that the sighting tube remains horizontal while you are reading heights.

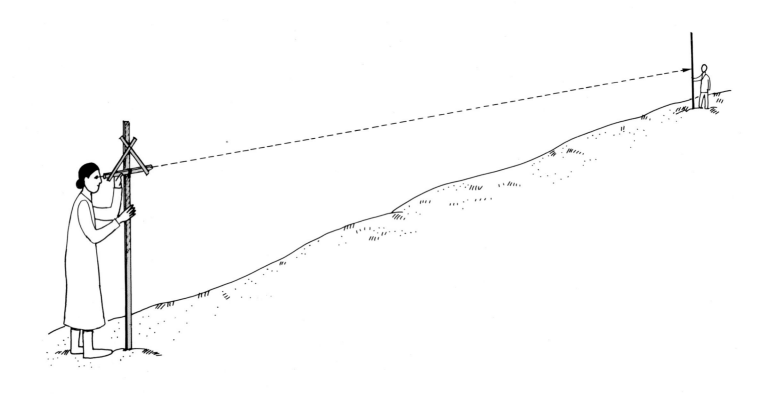

Making your own bamboo sighting level

1.   Get a piece of bamboo about 45 cm long and a few centimetres in diameter. Remove the inside membrane between its sections by drilling, or by driving a long object such as a metal rod through the tube.

2.   Across one end of the bamboo tube, glue two pieces of wire or thread at right angles to form a central sighting point.

3.   Cover the other end of the tube with tape. Waterproof plastic or electrical tape is best. Pierce the tape at its **centre** with a small nail, to make a sighting hole. When you use the level, you will look through this hole and read the measurement at the point where the two threads cross.

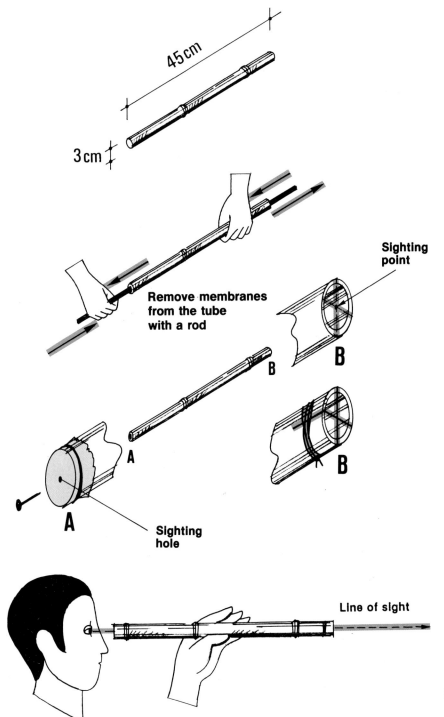

45 cm

3 cm

Remove membranes from the tube with a rod

Sighting point

B

B

B

A

A

A

Sighting hole

Line of sight

4. Place a small weight on the bamboo tube which can be moved along the tube to balance it. A hose clip makes a good balance, and it can be tightened to keep it in place once the tube has been horizontally adjusted.

5. Lash two 45 cm wood strips to opposite sides of the tube, near the ends, so that they form a triangle with the bamboo tube.

6. Drill a hole through each wooden strip at a point about 7 cm from the top.

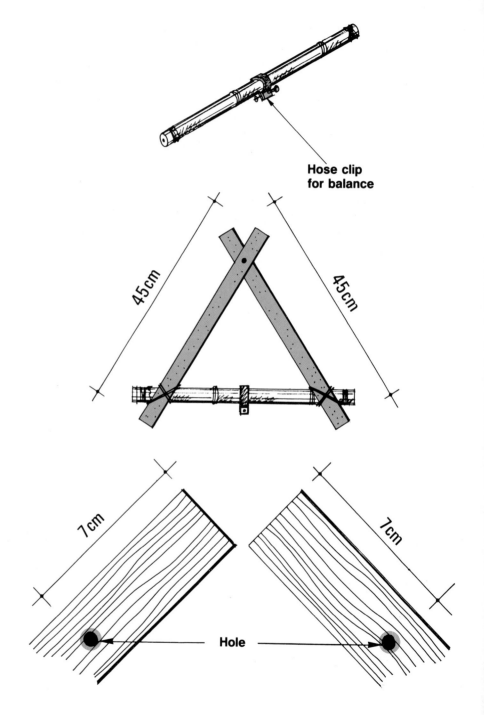

Hose clip
for balance

45 cm

45 cm

7 cm

7 cm

Hole

7. Get a 2 m vertical staff, and drill a hole through it near the top; the triangle sighting device will hang from this.

8. To allow the triangle to move freely, place small blocks of wood or short segments of bamboo between the wooden strips of the triangle where they cross at the top and between the back of the triangle at this point and the vertical staff.

9. Loosely bolt the triangle, through the wooden blocks or bamboo segments, to the hole in the vertical staff. **The sighting line should be exactly 1.50 m from the ground**. This height is convenient for both calculations and sighting. With the bamboo sighting tube perpendicular to the support staff, measure the vertical distance from the centre of the tube to the bottom of the staff. Mark a reference line 1.50 m below the line of sight.

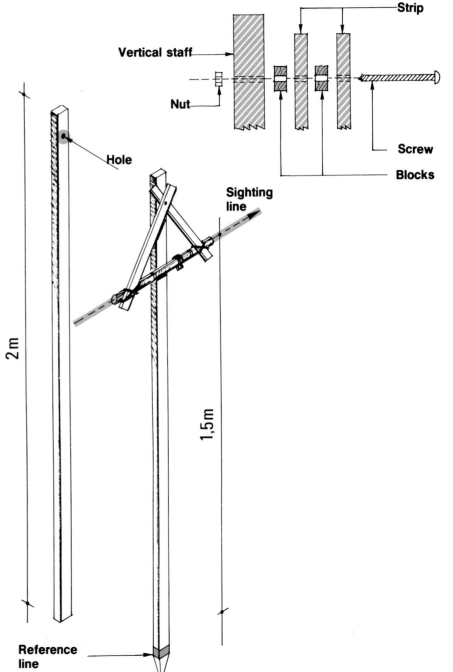

Adjusting the bamboo sighting level

10. Place the bamboo sighting level close to a 2 m measuring scale or levelling staff. Read the height on the scale by sighting at it through the small hole and reading the number that lines up with the crossed threads.

11. Move the measuring scale to a point which is 15 m away and at the same level. Check that this point is at the same level (with a straight-edge level, for example, see Section 51). Sight again through the bamboo tube and read the height on the scale to see if it is the same as before.

12. Check to see that the triangle is hanging freely by moving it with your finger. Let the triangle come to a stop and check the reading through the bamboo tube again to see if the result is the same.

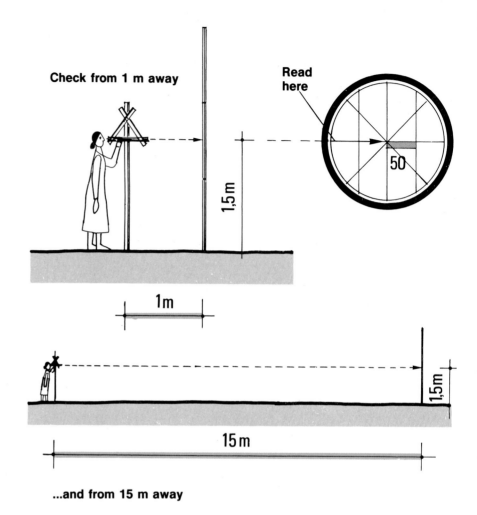

Check from 1 m away

Read here

50

1,5 m

1m

1,5 m

15 m

...and from 15 m away

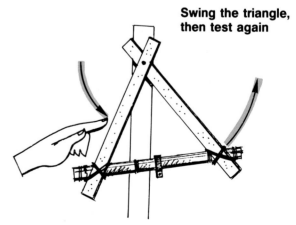

Swing the triangle, then test again

13. If the reading at the 15 m point is not the same as the reading from the point where the bamboo level and measuring scale were side by side, adjust the balance weight on the bamboo tube slightly. Move the weight towards the rear of the tube if the 15 m reading is lower; move it **forward** if the reading is higher.

14. Again place the bamboo sighting level and the measuring scale or staff side by side, and take a new reading.

15. Move the scale or staff 15 m away, and check this reading.

16. Repeat this process (see above, steps 10-15) until the two readings are the same.

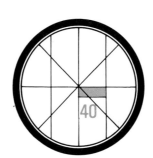 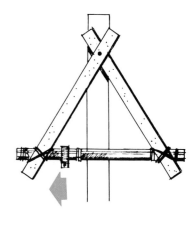

If the reading is low, move the weight back

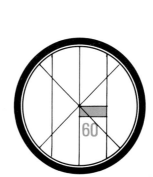 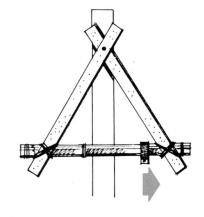

If the reading is high, move the weight forward

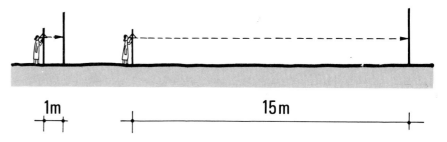

1m

15m

Keep testing until the sighting-tube is balanced   265

Note 1: If there is a small difference between the two readings after several repetitions, it may be caused by a slight difference in level between the two points. Interchange the positions of the bamboo level and the scale, putting the level at the 15 m point and the measuring scale at the 0 m point. Take another reading. Divide the difference in the readings by 2. Then, using this figure, make the bamboo tube horizontal by moving the balance weight along it.

Note 2: when the reading taken from 15 m away is **within 2 cm** of the reading you took with the bamboo level and the scale side by side, your bamboo level is accurate enough.

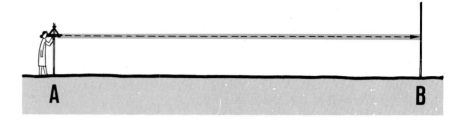

If the reading is still different

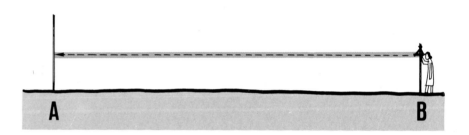

...exchange the positions of the level and the staff

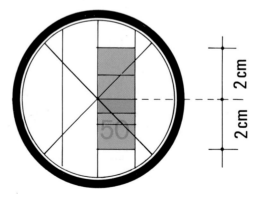

A variation up to 2 cm is acceptable

Using the bamboo sighting level for levelling

17. To use the bamboo sighting level you will need to work in **a team of two people**. The distance you can survey each time depends on how far away you can read the levelling staff graduation (usually 20 m, at the most).

18. You can level in either one or two directions, as described below.

Note: the bamboo level should be placed at each station so that the sighting line is 1.50 m above ground level.

Levelling in one direction only

19. Place the bamboo sighting level in a vertical position at point A, the beginning of the line you are surveying.

20. Your assistant should walk ahead 15 to 20 m along the line and place the levelling staff in a vertical position at point B, and mark point B with a stake.

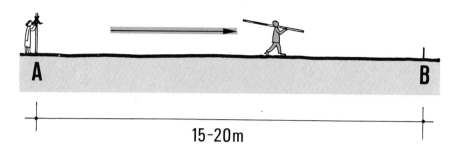

21. Take a reading on the levelling staff, from point A to point B and note it down. Then move forward to point B, and set up the bamboo sighting level where the stake was.

22. Your assistant should walk ahead another 15 to 20 m along the line. There the levelling staff is placed in a vertical position at point C which is marked with another stake.

23. Now take a reading on the levelling staff from point B to point C.

24. Repeat this procedure (see steps 22-23) until you have surveyed the entire line.

25. Carefully note down all the readings in a table and calculate the heights of the various points, if you need them (see Section 52, and Section 81 in Book 2).

26. When you have reached the end of the line you are surveying, you can also calculate the total height difference between the starting and the finishing points (see Section 52).

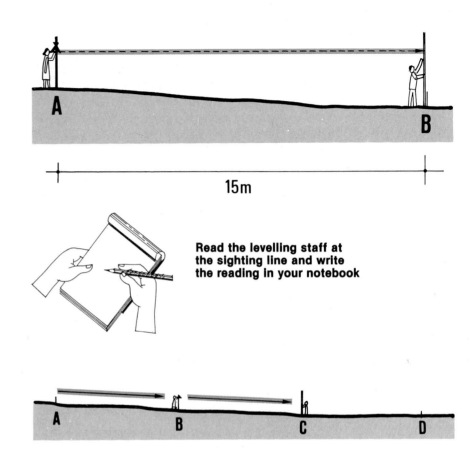

15m

Read the levelling staff at the sighting line and write the reading in your notebook

Move to the next station

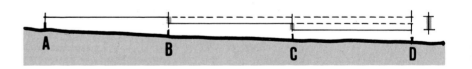

Add up all the readings

Levelling in two directions

27. You can measure two lengths of a line from a central point by sighting with the bamboo level in two directions. This system gives you two readings for each point except the first and the last. By comparing the **forward reading (FR)** and the **back reading (BR)** you can check the accuracy of your work.

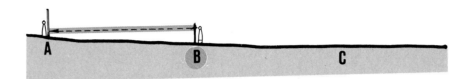

28. Your assistant should place the levelling staff in a vertical position at the starting point A on the line you need to survey.

29. Walk ahead 15 to 20 m along the line and place the bamboo level at point B. From there, take a **back reading (BR)** from point B to point A.

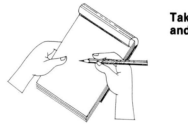

Take a back reading and write it down

30. Your assistant should then pace this distance to you, and then pace the same distance **past** you to the next point (C) ahead, where the levelling staff is placed.

31. Turn the bamboo level around at point B and take a **forward reading (FR)** from point B to point C.

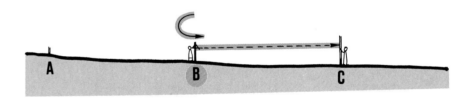

Turn the level around and take a forward reading, then write it down

32. Repeat this process until you have surveyed the entire line.

33. Note down all your readings in a table and calculate the height differences between the surveyed points (see Section 81).

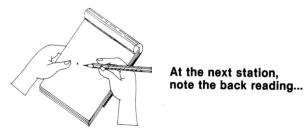

At the next station, note the back reading...

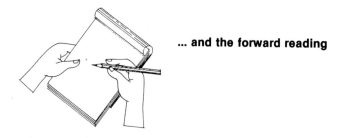

... and the forward reading

57 How to use the hand level

1. The hand level is a ready-made tool for quickly finding differences in level. Its range in the field should not exceed 15 m. You may be able to borrow a hand level from a local survey station or buy one from a hardware store. The hand level is made up of a **sighting channel, a spirit level** and a **mirror**. The mirror allows you to take a reading and, at the same time, check to see that the **sighting line*** is horizontal.

Using the hand level for levelling

2. The directions for using the hand level are the same as those given for the bamboo sighting level (see Section 57), except that:

- you can use it held in your hand;
- **the height of the sighting line** is the vertical distance from the ground to your eye level; and
- **the bubble** of the spirit level must be centred while you take the reading.

Note: you can have greater accuracy if you rest the hand level on the top of a wooden pole of convenient height. In this case, the height of the pole becomes the height of the sighting line.

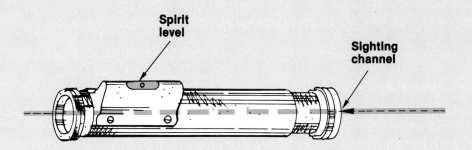

Spirit level

Sighting channel

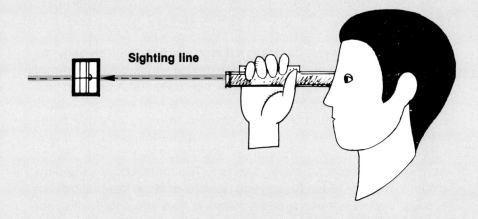

Sighting line

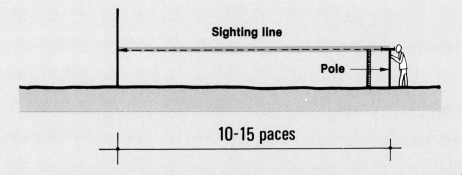

Sighting line

Pole

10-15 paces

58   How to use the surveyor's level and theodolite

1.   For very accurate levelling over long distances, surveyors use modern instruments called **surveyor's levels** and **theodolites**. These instruments are expensive and can be damaged easily. Only skilled personnel should operate, adjust and repair them.

2.   To level survey a small farm you will not usually need the high accuracy of these instruments, and you may use cheaper devices. You have learned about these in earlier sections. However, either a **surveyor's level** or a **theodolite** may be available for your survey. Both should be used with a **modern** levelling staff (see Section 50, step 8) to give the greatest accuracy. The levelling staff is set vertically into the ground so that **its graduation marks are upside down**, since the sighting devices on the surveyor's instruments invert the images, making them appear upside down.

3.   A **surveyor's level** is basically a **telescope, fitted with cross-wires** for sighting, and attached to a **levelling device** which is mounted on a **tripod** (a support with three legs). In older instruments, the horizontality of the sighting line was adjusted with a sensitive spirit level and fine-threaded adjusting screws. In more recently made instruments (known as self-levelling or **automatic levels**), the line of sight is automatically brought to the horizontal, which makes surveying operations much easier. The telescope **magnifies** far-away objects, which means you can observe the graduation on a levelling staff at a much greater distance than you could with your ordinary eyesight.

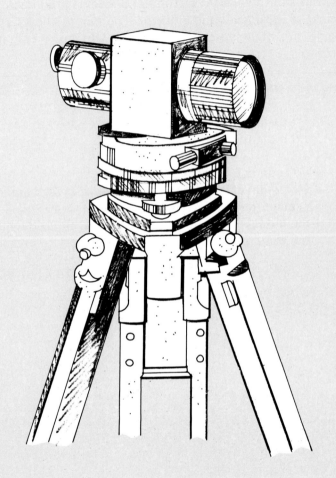

4.   **Theodolites** are commonly used to measure **horizontal** angles (see Section 35) and **vertical** angles (see Section 47). They can also be used to measure **height differences**.

5.   Most surveyor's levels and theodolites are equipped with **stadia hairs**. These allow you to determine distances during level surveys (see Section 28).

6.   Height differences are measured by using the **horizontal sighting line** as a reference, as described for the bamboo sighting level (see Section 57). These differences are recorded and calculated as explained in Section 81 in Book 2. Very long lines can be surveyed much faster, without measuring as many intermediate stations.

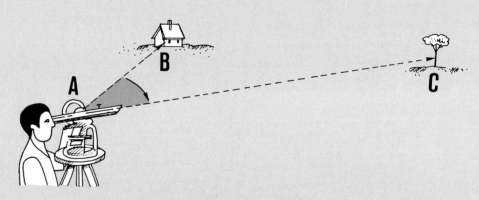

Measuring horizontal angles

6 MEASURING HEIGHT DIFFERENCES – PART 2

60 Introduction

1. In the previous chapter, you learned how to make several types of simple levels. You also learned how to use them in the field for **levelling**.

2. These levels can also be used for finding and marking on the ground **all points at the same height**, such as the points along the centre-line of a future water supply canal. In this case, the height differences between the various points of the line would be made equal to zero. These points make up a **contour line**. This particular type of level survey is called **contouring**. There are some simple levels which can be used for contouring. These will be described in the following Sections 61-65. How to use other levels and slope measuring devices for contouring will be explained in Sections 66-68.

3. In this chapter you will also learn how to use the levelling devices already described to set lines of slope with a gradient (see Section 69).

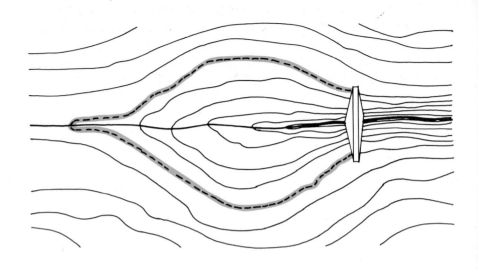

Contour lines

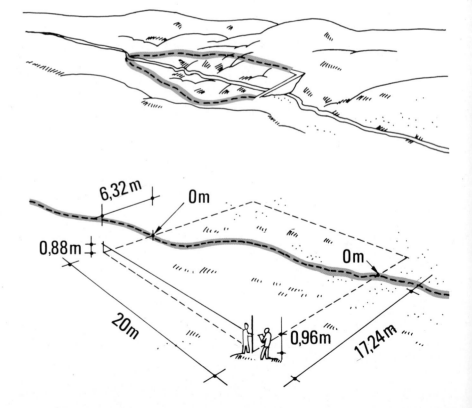

4. There are several good **ways of laying out contour lines**. Each of these methods is fully explained in the next sections. **Table 8** will also help you choose the method best suited to your needs. Later, in Section 83, you will learn how to lay out contours in the field, and in Section 94, you will learn how to map the results of your field survey.

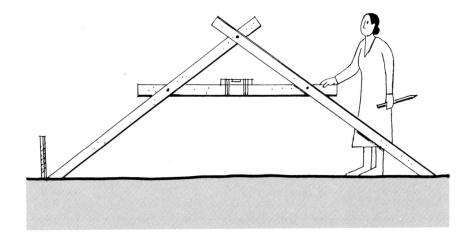

TABLE 8
Contour levelling methods

CONTOURING LEVELS

Section [1]	Method [2]	Distance, m	Accuracy	Remarks	People, equipment
62 *	*A-frame*	4	Medium	Awkward to transport	1 or 2 people, mason's level
63 *	*A-frame, plumb line*	4	Medium to high	Fast to use	1 or 2 people, plumb line
64 *	*H-frame water level*	2.5	Medium to high	Awkward but quick Avoid water loss	2 people
65 *	*Semi-circular water level*	100	Medium	Faster for longer distances Avoid water loss	2 people, target levelling staff

NON-SIGHTING LEVELS (see also Table 7)

Section [1]	Method [2]	Distance, m	Accuracy	Remarks	People, equipment
66 *	*Straight-edge level*	2.5 to 3	Medium to high	Easy transport Fast	1 person, mason's level
66 **	*Line level*	20	Medium	Very easy to transport Quick to operate Useful on rough ground	3 people, mason's level 2 measuring scales
66 *	*Flexible tube water level*	10 to 15	High to very high	Awkward to transport Very quick Avoid water loss	2 people, 2 measuring scales

SIGHTING LEVELS (see also Table 7)

Section [1]	Method [2]	Distance, m	Accuracy	Remarks	People, equipment
67 **	*Bamboo level*	15 to 20	Low to medium	Greatly affected by wind	2 people, 1 levelling staff
67 *	Hand level	10 to 15	Low	Rough, fast	2 people, 1 levelling staff, hand level
67 ***	Surveyor's level	more than 100	Very high	Expensive, delicate	2 people, 1 levelling staff, surveyor's level

SLOPE MEASURING DEVICES (see also Table 6)

Section [1]	Method [2]	Distance, m	Accuracy	Remarks	People, equipment
68 **	*Clinometer*, clisimeter	10 to 15	Low to high	See Table 6	2 people, levelling staff

[1] * Simple   ** more difficult   *** most difficult

[2] *In italics*, equipment you can make yourself

5. In nearly all levelling instruments, horizontality is shown by a **spirit level**. This is a small level, usually made of an elongated or circular glass tube; the tube is nearly filled with a liquid (usually spirit), leaving enough space to form an **air bubble. In the elongated spirit level**, a point near the middle of the tube is selected as the **zero-point**, and clearly marked. Graduations may be added on either side of this point. **In the circular spirit level**, the zero-point lines up with the centre of the level, and is clearly marked by a small circle. When the air bubble is at the zero-point, the level is horizontal.

Circular spirit level

Air bubble   **Liquid**

Elongated spirit level

61   How to use the mason's level

The mason's level is a simple tool often used during building operations. You have learned how to use the mason's level to **set out horizontal lines** when you measure short distances on sloping ground (see Section 21) and when you determine height differences (see Sections 51-52).

What is a mason's level?

1.   Usually a mason's level consists of a rectangular wooden casing with a small **spirit level** mounted in one of its narrow faces. The mason's level can also be made of metal. Glass spirit levels are highly breakable, and should be handled very carefully.

2.   The casing varies in length. **As the length increases, the accuracy improves**. The cheapest mason's levels are relatively short, about 25 m long. They are generally available from hardware stores.

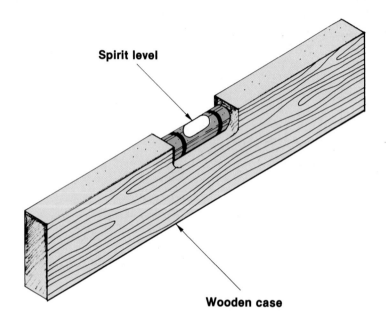

Spirit level

Wooden case

Mason's level

Using a mason's level to check horizontality

3. When a mason's level is horizontal, the **bubble of the spirit level** lies exactly at its zero-point.

4. When the air bubble moves away from the zero-point, it shows that the level is no longer horizontal. There is either an uphill or a downhill slope.

Note: the direction in which the bubble moves always indicates the direction of the **highest point on the slope.**

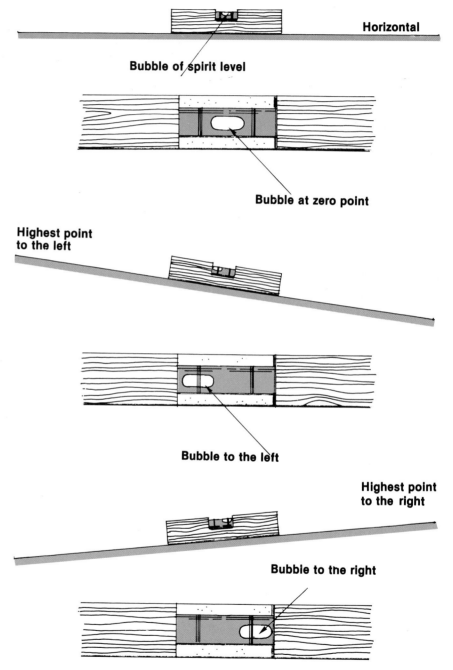

Horizontal

Bubble of spirit level

Bubble at zero point

Highest point to the left

Bubble to the left

Highest point to the right

Bubble to the right

62   How to use the A-frame level

A simple device for contouring can be made from three pieces of wood and a mason's level. This device works on the same principle as the **straight-edge level** (see Section 51), but it is easier and faster to use.

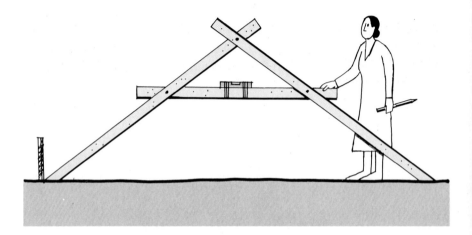

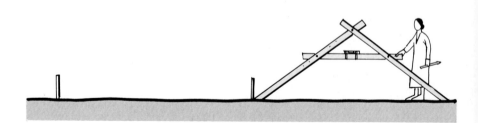

Making your own A-frame level

1.   Get three pieces of soft wood, all at least 2 × 6 cm thick; two of the three pieces should be about 2.80 m long, and the other about 2 m long. The A-frame made of these will be about 1.70 m high by 4 m long -- small enough to handle easily.

2.   Attach the two 2.80 long **leg-pieces** about 30 cm down from their tops by drilling a hole through the centre of each piece and bolting them loosely together. Adjust the legs until they are 4 m apart at the bottom.

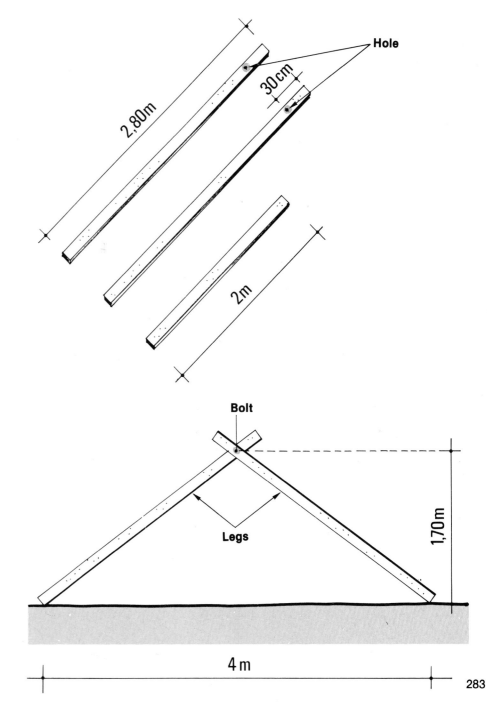

3. Measure up 1.60 m from the bottom of each leg and loosely attach the 2 m **cross-piece** by drilling and bolting it to the legs. The cross-piece should be about 1 m above the ground.

4. Cut the bottom of the legs level, so that they rest evenly on the ground when the A-frame is upright. To do this, stand the A-frame upright on its legs and place a long straight piece of wood so that it touches both legs at the base. Make a mark along the legs, level with the top of this piece of wood, and cut the legs at the mark.

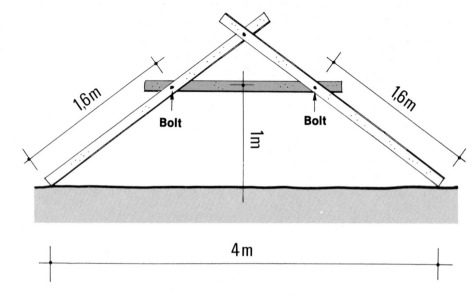

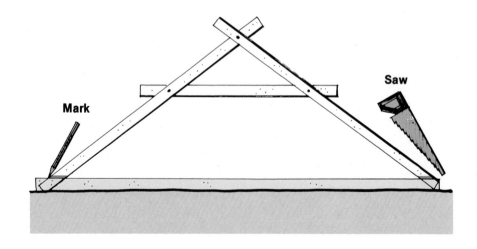

Adjusting the A-frame level

5.   Place the A-frame upright, resting the legs on two points which are at **exactly the same level**. Put the **mason's level** (see Section 61) on the middle part of the cross-piece and check to see if it is **horizontal**. If it is not, adjust it by moving the cross-piece slightly, or by cutting a little more off one of the legs. When the cross-piece is horizontal, tighten all the bolts on the A-frame.

6.   To check the horizontality, turn the A-frame around and, with the mason's level, check to see that the cross-piece is still horizontal.

7.   Using light string, lash the **mason's level** securely to the cross-piece at its **mid-point**.

Check the level

Then reverse and check again

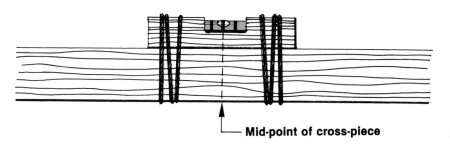

Mid-point of cross-piece

285

Using the A-frame for contouring

8. With a marking pin, mark point A where you will begin contouring. Place one leg of the A-frame at this point. Move the other leg uphill or downhill until the **mason's level** shows a horizontal position. At this point place another marking pin B.

9. Move the A-frame up to the second point B. Find the next horizontal point C and mark it.

10. Repeat this process until you have plotted the **length of the contour line AE.**

Adjust the forward leg until the level is horizontal

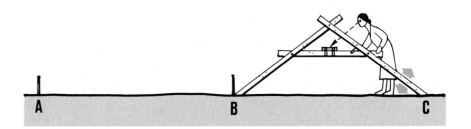

Repeat at each station

Contour line AE

286

63   How to use the A-frame and plumb-line level

The A-frame and plumb-line level is a simple device very similar to the A-frame, except that the mason's level is replaced by a plumb-line. The device is used in the same way as the standard A-frame for contouring (see Section 62).

Making your A-frame and plumb-line level

1.   Construct an **A-frame** as described above (see Section 62, steps 1-4).

2.   Screw a small hook, or drive a nail, into the frame near its summit.

3.   Attach **a plumb-line** (see Section 48) to the hook or nail. The plumb-line should be long enough for the plumb to reach below the cross-piece of the frame.

Note: the taller the frame is, the more sensitive the level will be to differences in height. The dimensions given in Section 62 provide a good **average accuracy**, usually better than 32 cm over 100 m.

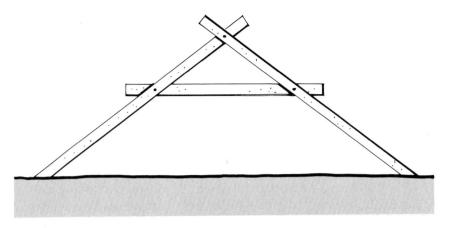

A-frame

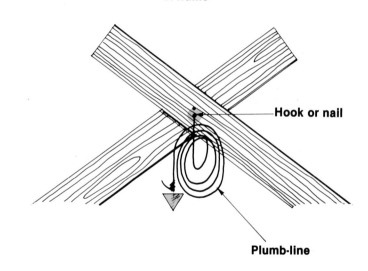

Hook or nail

Plumb-line

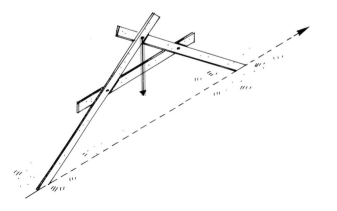

287

Adjusting your A-frame and plumb-line level

4.   Place the A-frame upright with its legs resting on two points which are at exactly the same level.

5.   When the plumb-line comes to rest, lightly mark the position of its string on the top side of the cross-piece of the A-frame.

6.   Place the A-frame the other way around, so that its legs are reversed on the same horizontal points. When the plumb-line comes to a stop, lightly mark the position of the string on the cross-piece.

7.   Make a **permanent mark** on the front side of the cross-piece at the precise mid-point between the two marks. This shows where the legs of the A-frame are exactly level.

Note: to improve measurements **in windy weather**, slow the movement of the plumb-line letting it rub slightly against the cross-piece of the A-frame.

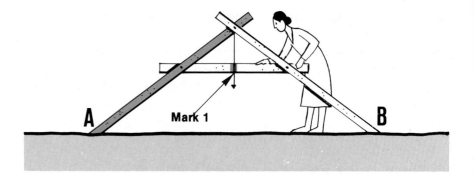

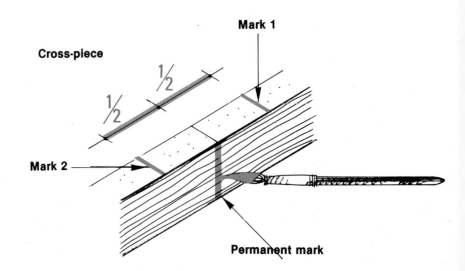

64   How to use the H-frame water level

The H-frame water level is a simple device made of a light wooden frame and some clear plastic piping, which is partly filled with water. Like the flexible-tube water level (see Section 53), it is based on the principle that, under atmospheric pressure, the free surfaces of interconnected water columns will reach equal heights, which follow a horizontal line.

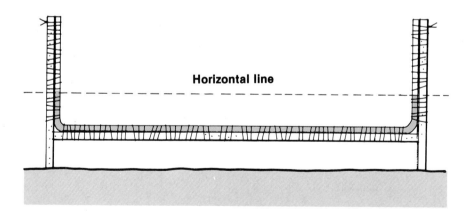

Horizontal line

Making your H-frame water level

1. Get two 5 × 5 cm thick pieces of soft wood 1 m long, and one 5 × 5 cm piece 2.5 m long. Join the three pieces of wood together to form an "H" shape, using strong nails or bolts. **The horizontal piece** of the frame should be about 20 cm above ground level. The two **upright legs** should make 90° angles with the horizontal piece. Check this.

2. Get 3.90 of clear, non-reinforced **plastic tubing** with an inside diameter of about 1.2 cm. Using soft wire or string, secure it to the upper face of the horizontal piece and to the inside faces of the two vertical pieces. Tie or bind the plastic tube tightly to the wooden pieces, but be careful not to pinch the tube.

Note: if you do not have enough clear plastic tubing, use about 1.90 m of dark rubber or plastic piping or metal water piping, and two 1 m lengths of clear plastic tubing. Connect one length of clear tubing to each end of the dark piping with a hose clip. Then tie the dark piping to the horizontal piece of the H-frame, and the clear lengths of tube to the two vertical pieces.

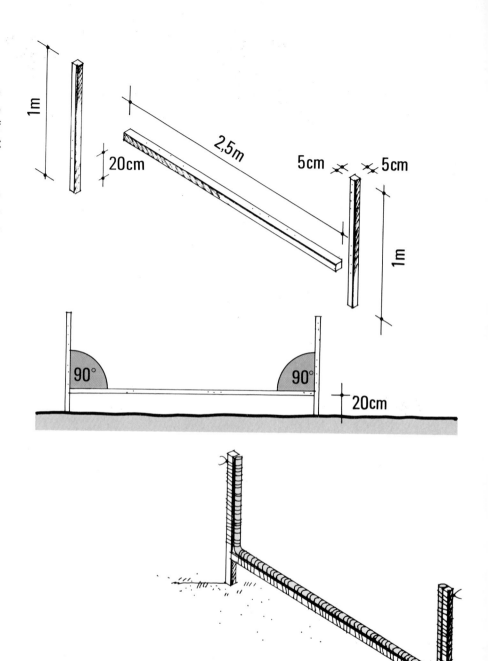

3. Pour water into the tubing until the level reaches about **halfway** up each vertical section, making sure to get rid of any air bubbles. Put a **cork stopper** in each tube-end to prevent water losses during transportation.

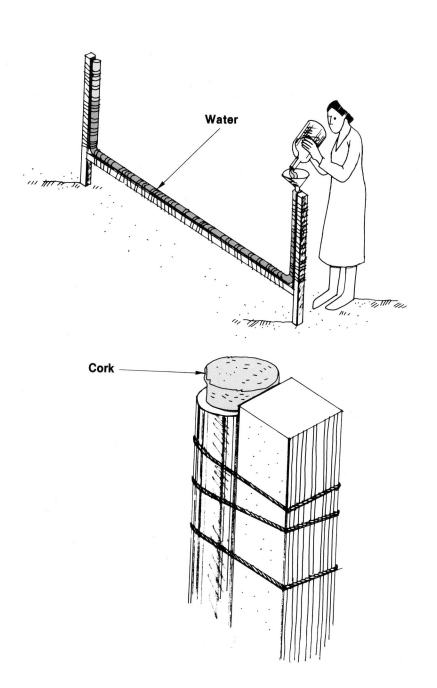

Adjusting your H-frame water level

4.   With the help of an assistant, place the H-frame upright, with its legs resting on two points which are at exactly the same level.

5.   **Remove the two stoppers** from the tube ends and look at the water level in each tube from the side. You and your assistant should then make a light mark on each vertical leg, level with the water level in the tubes.

6.   Turn the **H-frame around** and place its legs, reversed, on the same points.

7.   Again, lightly mark the water level on each vertical leg.

8.   Make a **permanent mark** on each leg at the precise mid-point between the two previous marks. When the water is at this level in the tubes, it indicates horizontality.

9.   **Replace the stoppers** for transportation.

Note: it is best to check this adjustment before each contouring survey. If any water has been spilled from the tubing, you should adjust the device by adding water as necessary.

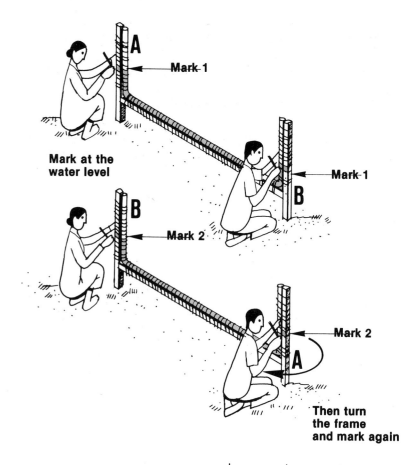

Mark at the water level

Then turn the frame and mark again

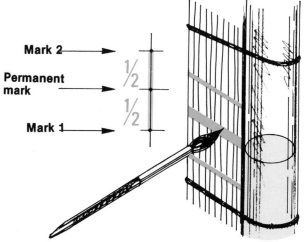

292

Using your H-frame water level for contouring

10.   Place the rear leg of the H-frame at the starting point A.

11.   Remove the stoppers from the tube ends.

12.   Move the forward leg uphill or downhill until the top water level reaches the permanent mark you have made on the leg.

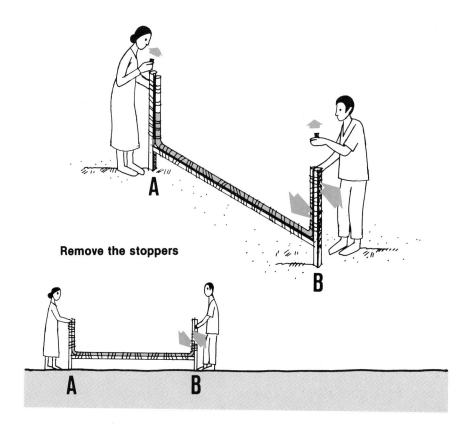

Remove the stoppers

Adjust the forward leg

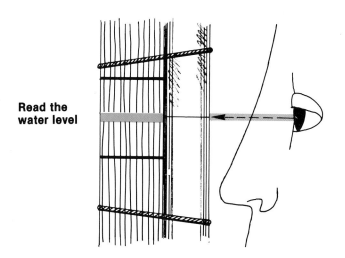

Read the water level

13.   Mark the position of the forward leg at B with a peg and replace the stoppers in the tube ends.

14.   Move the frame forward, place the rear leg at the marked point B, and repeat the previous procedure. Continue in this way until you reach the end of the contour line AE.

Note: it is easier to work with an assistant, who can move the forward leg until he or she finds the horizontal level. Then you can check that the water level on the rear leg also lies opposite the permanent mark.

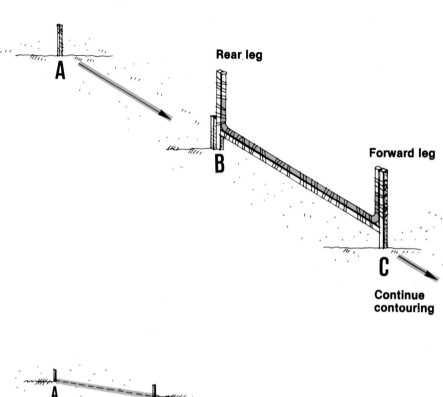

Rear leg

Forward leg

Continue contouring

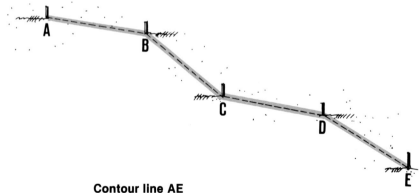

Contour line AE

65   How to use the semi-circular water level

The semi-circular water level is a simple device based on the same principle as the H-frame water level. Its main advantage is that you can use it on longer distances without moving it. You need only several small pieces of wood and a short piece of clear plastic tubing to make it, but it is a little more difficult to build than the H-frame.

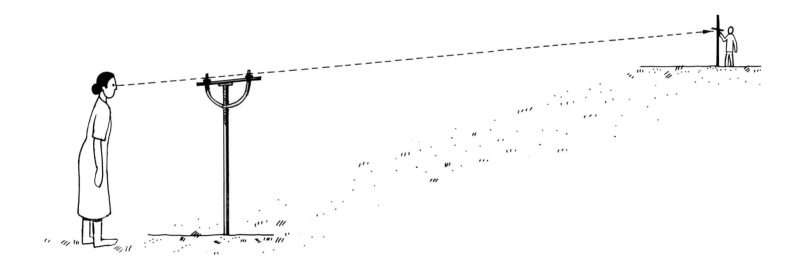

Making your semi-circular water level

1.   Get a 1 × 10 cm piece of wood 60 cm long and drill a hole through each end of it from the 10 × 60 cm face. These holes should be just wide enough to hold the plastic tube (see step 5).

2.   Drill a small hole in the centre of the piece of wood.

3.   Prepare two **wooden discs** with a diameter of 10 cm, and drill small holes in their centres.

4.   Nail or screw one of these discs under the centre of the piece of wood, **aligning the hole** in the centre of this piece with the hole in the disc. Do not block the hole.

5.   Get a piece of **clear plastic tube** about 80 cm long and 1 to 1.5 cm in diameter. Pass the ends of the tube from below through the holes in the ends of the piece of wood so that the tube forms a semi-circle on the side where the disc is. The two ends of the tube should extend above the piece of wood by about 10 cm. Keep the tubing in place by putting a **hose clip** just at the point where the tubing passes through the hole in the board. Tighten the clip so that the tube does not slip, but be careful not to pinch the tube. The hose clip will keep the tube in place, since it is bigger than the hole.

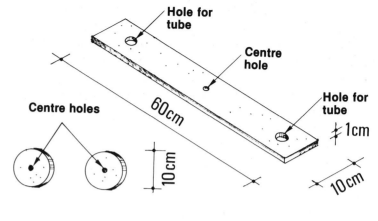

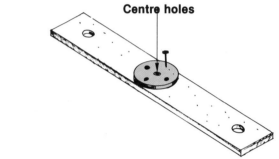

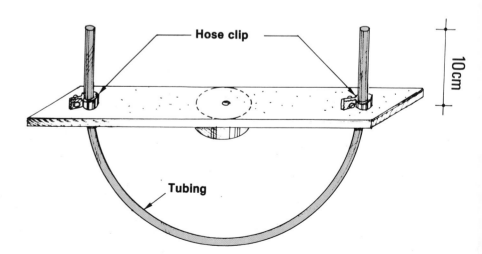

6. Now make the supporting leg. Get a pole 5 cm in diameter and 1.40 m long. Find the **centre point** of one end. Then take the second disc you prepared, and loosely nail it to the pole so that its centre hole is over the centre of the pole.

7. Attach the semi-circular level you made in step 5 to the supporting leg. Use a strong screw, and align the central holes of the wooden discs carefully. Do not tighten the screw too much. You must be able to turn the semi-circular level around. The flexible tube will be off to one side of the pole.

8. Place the device upright on its support and fill the plastic tube with water. The level of the water should reach about 4 to 5 cm from each end of the tube. Place a **stopper** in each end of the tube to prevent water loss during transportation.

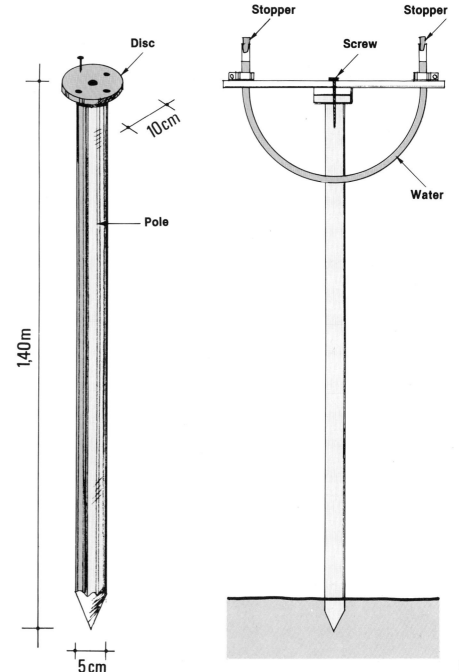

297

Using your semi-circular water level for contouring

9. At the starting point A of the contour you need to level, ask your assistant to place a levelling staff in a vertical position. Since you are contouring with a sighting level which does not include a telescope, you should use **a target levelling staff**.

You can easily make one. Get a straight wooden stick, a piece of bamboo, or a maize stalk 2 m long. Get another pole or stick 50 cm long, and attach it to the first one with string, to form a cross. The location of the point where you attach the 50 cm pole, called the **target**, depends on the contour you are levelling.

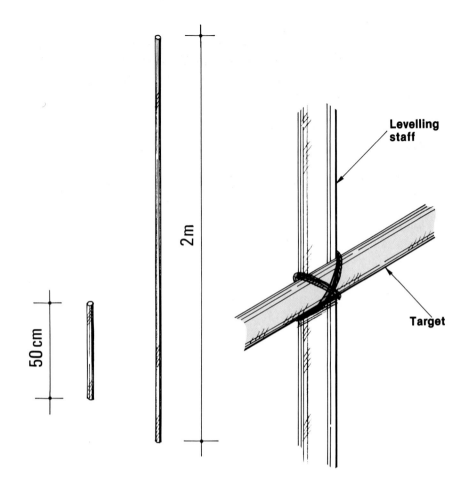

Levelling staff

Target

2 m

50 cm

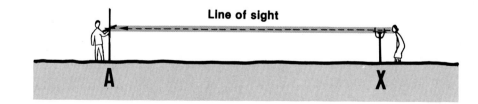

Line of sight

A   X

10. To station the semi-circular water level, drive its support vertically into the ground at a central point from which you will be able to survey about 100 m of the contour line. Remove the stoppers from the ends of the plastic tube.

11. Standing about 1 m behind the semi-circular water level, rotate its upper part and sight along a line which joins the two water-surface levels in the plastic tube to the levelling staff. Signal to your assistant to adjust the target of the levelling staff up or down until it is exactly on the sighting line. Then, ask your assistant to tie the target firmly at that height.

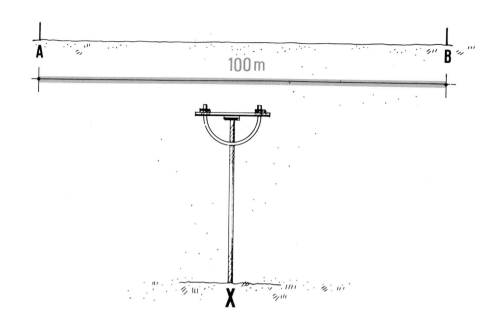

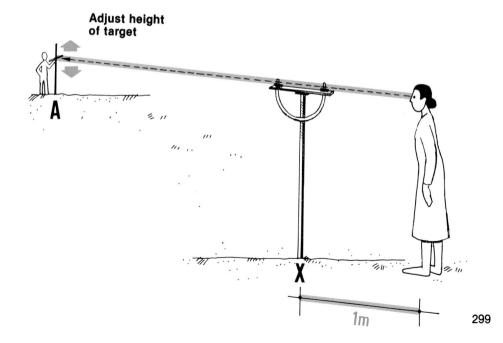

12. Your assistant will then mark the starting point A with a stake, and walk about 10 m away, where he will place the levelling staff in a vertical position.

13. Rotate the upper part of the water level until you can sight again at the cross on the staff. Signal to your assistant to move the levelling staff uphill or downhill until the fixed target lines up with the sighting line. He or she will then mark this point B with a stake.

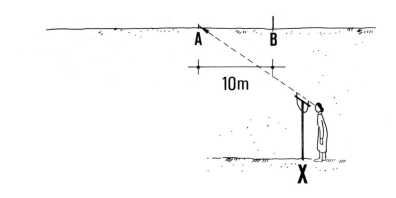

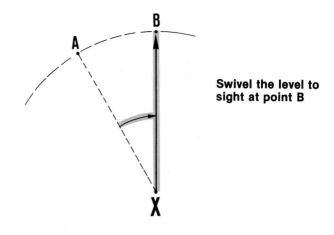

Swivel the level to sight at point B

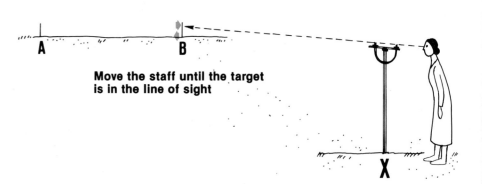

Move the staff until the target is in the line of sight

14. You may keep on levelling points on the same contour AG from one central station X for about 100 m. To continue the same contour line, leave the target levelling staff at point G, and move the level to **a new central station Y**. Adjust the **height of the target** and go on levelling contour GM from station Y.

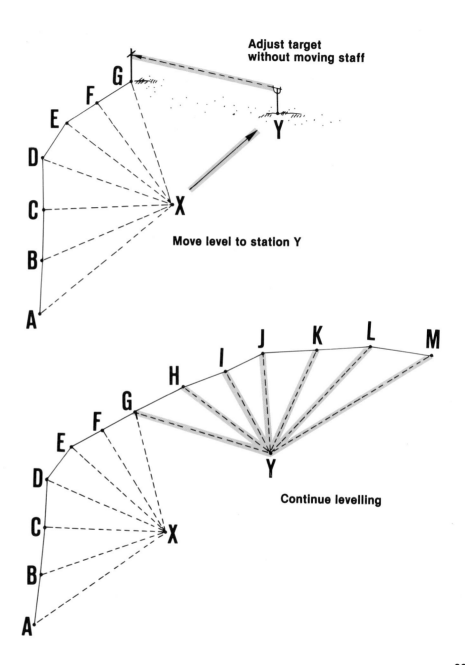

Adjust target without moving staff

Move level to station Y

Continue levelling

Note: you may also want to find contours with a fixed height difference (see Section 83 in Book 2), for example, every 0.20 m. To do this, you will keep working from the same station, but change the height of the target on the levelling staff. When you reach point G, have your assistant lower the target by 20 cm. He then walks up the hill along line XGH until the target is level with your line of sight, marking point H on the next contour. Continue the second contour line HN by finding point I on line XFI, point J on line EXJ, and so on. If the distance is short enough for you to see clearly, you may lower the target again to set a third contour line from the same station.

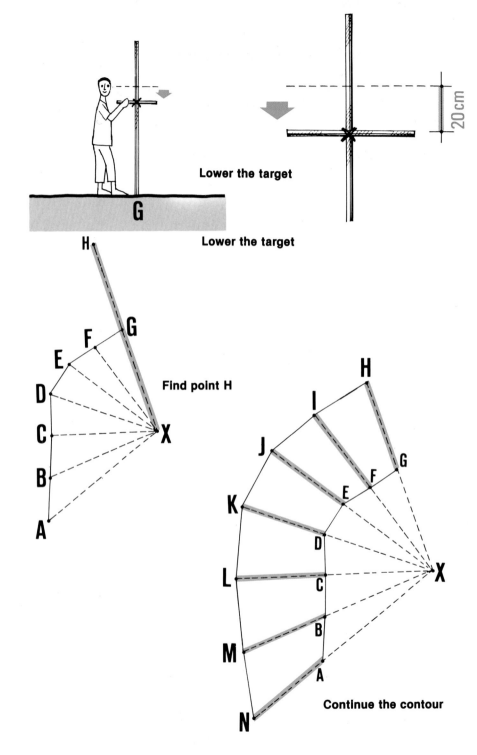

Lower the target

20 cm

Lower the target

Find point H

Continue the contour

66 How to contour with non-sighting levels

1.   In Sections 51 to 53, you learned how to use non-sighting levels to measure differences in height. These devices can also be used for contouring.

Using the straight-edge level for contouring

In Section 51, you learned about the straight-edge level. For contouring, use it in the following way (steps 2 to 7).

2.   Mark the point A where you will begin contouring with a stake. Place one end of the straight-edge at this point, and move the other end uphill or downhill until the mason's level shows horizontality. Mark this point B with another stake.

3.   Move the straight-edge up to point B. Find the next horizontal point C as shown above and mark it with a stake.

4.   Continue this process until you have marked the length of the contour line.

Make the level horizontal

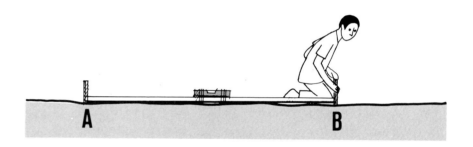

Mark point B

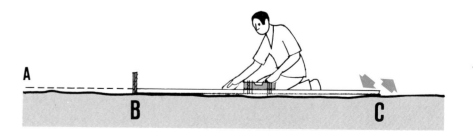

Continue the contour line

5. Mark the route of the contour line you have found by leaving a stake about every 10 m. If the contour curves, you may need to use more stakes.

6. If the surface of the ground is somewhat rough (i.e. covered with lumps of earth, stones or grass), it may help to use two bricks or wooden blocks of the **same height** to support the ends of the straight-edge while you are levelling.

7. If the surface of the ground is very rough or covered with dense grass, you can use two stakes under the ends of the straight-edge to lift it above ground level. Be sure that **both stakes are the same length**, and that you drive them into the ground to the same level. This way you can transfer the horizontal you find, which lines up with the top level of the stakes, to ground level without error.

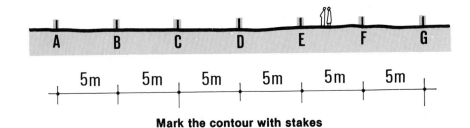

Mark the contour with stakes

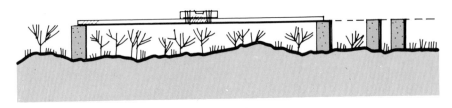

Use bricks

...or stakes to lift the level above obstacles

304

Using the line level for contouring

In Section 52, you learned how to make a line level. The line level is very efficient for contouring because it allows you to progress quickly, even on rough grass, and it is reasonably accurate (the maximum error is less than 6 cm per 20 m distance). Remember that you need three people to use the line level.

8. The rear person places one end-staff on the marked starting point A and keeps the cord on the 1 m graduation, for example. The front person, also keeping the cord on the same graduation, moves the second end-staff up or down the slope until **the centre person** signals that the mason's level is horizontal. The front person then marks the point B where the staff touches the ground.

9. The rear person walks to this marked point B while the other two people walk ahead until the cords are well stretched. The entire procedure is repeated and another point C of the contour line is marked.

10. This process is continued until you have marked the length of the contour line.

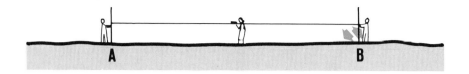

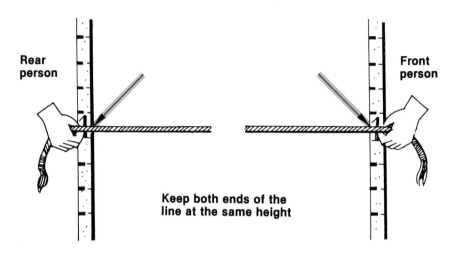

Rear person

Front person

Keep both ends of the line at the same height

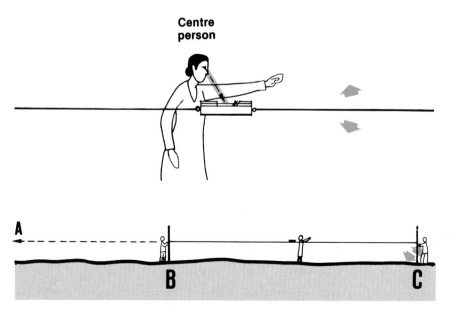

Centre person

305

Using the flexible-tube water level for contouring

In Section 53, you learned how to make the **flexible-tube water level**. You can contour quite quickly with this device even on rough ground, and it will give very good accuracy (the maximum error is about 1 cm per 10 m distance). You should be very careful not to lose water during the procedure. You will need an assistant for this method.

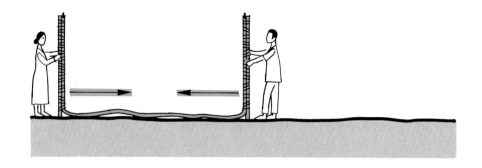

11.   Bring the two stand pipes together at the starting point A of the contour line, remove the stoppers, and mark the height of the water levels on each measuring scale. These heights should be the same.

12.   Replace the stoppers in the tube ends.

Put both stand pipes at point A and remove the stoppers

Mark the water level on both scales

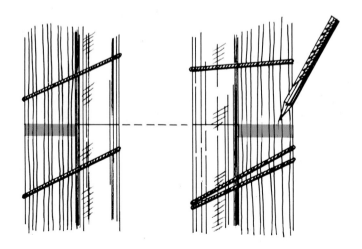

13.   Place your measuring scale at the starting point A of the contour line. Have your assistant walk ahead until the end of the hose is reached. Both of you remove the stoppers, and your assistant moves the scale up or down the slope until the water level is at the **marked height**. Check that the water level is at the marked height at your end, too. When it is, signal to your assistant to mark the location B of that scale with a stake. Replace both the stoppers.

14.   Both of you then move forward until you are standing at the point B where your assistant was standing, as marked with the stake. Have your assistant walk ahead until the end of the hose is reached. Repeat the procedure in step 13, and continue in the same way to the end of the contour line.

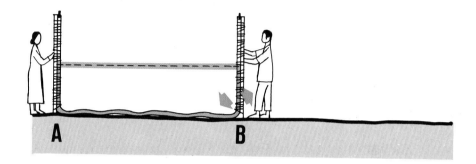

When the water level is at the mark on the standpipe, you have found point B

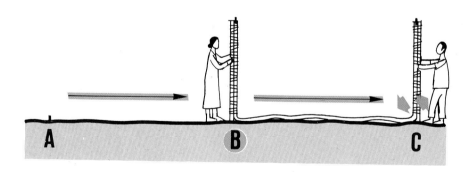

Begin the next measurement at point B

307

67 How to contour with sighting levels

1. In Sections 56 to 59, you learned how to use sighting levels to measure differences in heights. These devices can also be used for contouring.

Using the bamboo sighting level for contouring

2. You learned how to make and adjust a bamboo sighting level in Section 57. You and an assistant can use this level for contouring, as follows.

3. Place the bamboo sighting level next to a levelling staff and read the height on the scale by sighting through the tube.

4. Mark this height on the scale. You can use paint, or tie a piece of cord or a coloured rag at that height. You can also use the **target levelling staff** described in Section 65, and attach the target at that height.

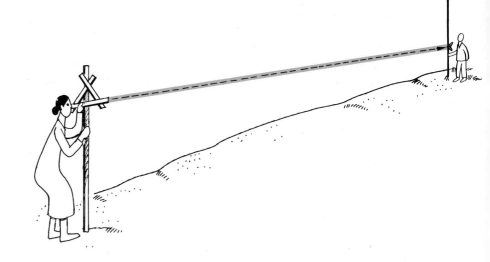

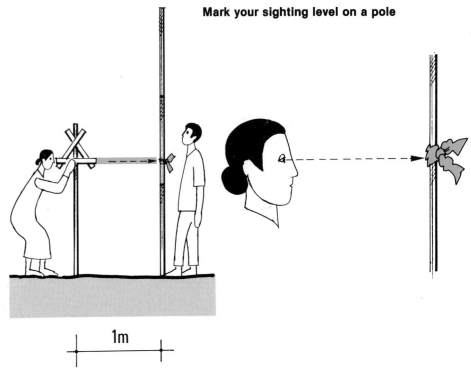

Mark your sighting level on a pole

1m

5. Place the bamboo level in a vertical position at A, the beginning of the contour line you want to plot.

6. Have your assistant, holding the levelling staff, walk 15 to 20 m ahead along an approximate horizontal line and place the staff vertically. Have him move it up or down the slope until you signal that the mark is lined up with the sighting line seen through the bamboo tube.

7. You will have to turn the bamboo level from left to right to see the mark on the measuring scale. Check frequently to make sure that the bamboo tube remains horizontal.

8. When you signal that you have sighted the mark, your assistant should mark the position B of the levelling staff with a stake.

9. Now move up to this stake B and place the bamboo level on that point in a vertical position.

10. Have your assistant walk another 15 to 20 m ahead with the levelling staff, and repeat steps 7-9.

11. Repeat this process until you have marked the entire length of the contour line.

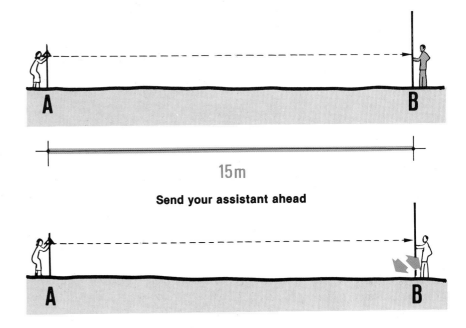

Send your assistant ahead

Have him move until you sight the mark

You may have to turn the level to see him

Proceed from point B

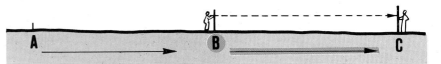

309

Using the hand level for contouring

12.   You can survey a contour line quickly by using a hand level (see Section 58), although this will not give the most accurate results. The method you use with the hand level is the same as that just described for the bamboo level, except that you should make the mark on the levelling staff at the height of the sighting line. The sighting line's height will either be at your eye-level, or at the height of the pole supporting the hand level (which is used to improve accuracy). The distance from one point to the next should not exceed 15 m.

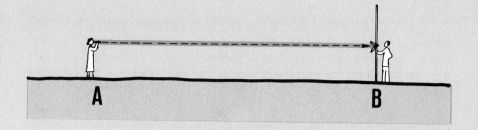

Use the hand level alone

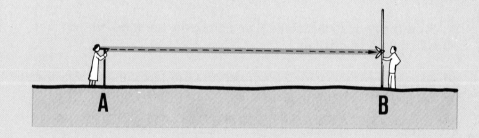

...or with a support staff

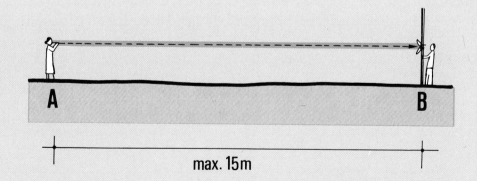

max. 15 m

Using the surveyor's level or the theodolite for contouring

13.   You can very quickly and accurately determine contour lines with a surveyor's level or a theodolite and **a precisely graduated levelling staff** (see Section 59).

14.   Since the **range of the telescope** on either of these devices is several hundred metres, you can reduce the number of stations. As you did with the semi-circular water level (see Section 65), you can survey several points from a single station. In open country, it is possible to use this method over long distances. In areas with forests, you might need to measure over shorter distances and to clear sighting lines.

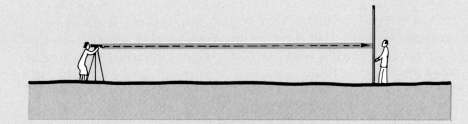

**Use the theodolite with
a graduated levelling staff**

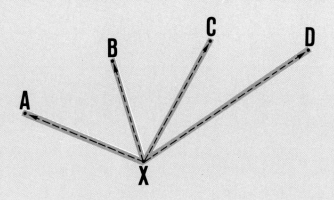

**You can survey several points
from a single station**

68   How to contour with slope-measuring devices

1.   In Sections 41 to 46, you learned how to use various types of clinometers to measure slopes. These devices can also be used for contouring because a contour line is defined as a line along which the **slope gradient equals zero**, see Section 83 in Book 2.

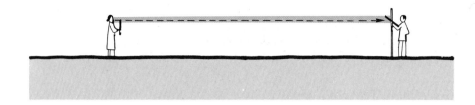

Sight at a target levelling staff

2.   When you contour with slope measuring devices, it is best to use a **target levelling staff**, such as the cross-shaped one described in Section 65. If you use such a staff, the target should be tightly fixed **at eye level.**

Note: if you do not have a levelling staff, you can use the **height of your assistant** as a **reference level*** instead.

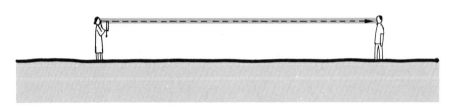

...or at your assistant

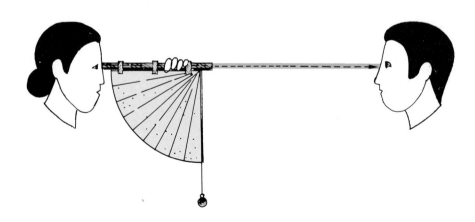

3. Your assistant, holding the levelling staff **vertically**, should stand about 10 to 15 m away from the starting point A of the contour line you want to plot. From this starting point, use the clinometer to sight at the levelling staff. Signal to your assistant to move the staff up or down the slope, until its target lines up with the **zero-graduation** of your clinometer. Have your assistant mark this ground point B and repeat the same procedure from it.

Note: if you are using a **clisimeter**, remember that you should use the **left scale** and make the sight of the levelling staff line up with its zero line (see Section 45).

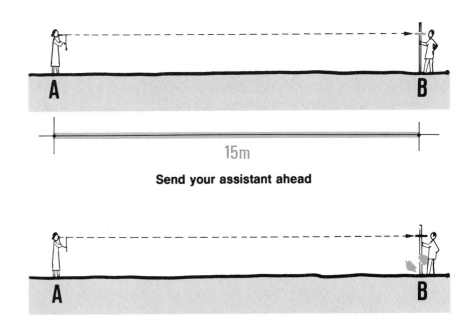

Send your assistant ahead

Have him move until you sight the mark

Proceed from point B

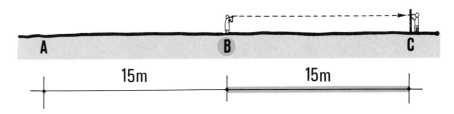

313

69 How to set graded lines of slope

1. **Graded lines of slope** are often used in fish-farms to assist **gravity*** in moving water. Water-supply canals and pipelines, as well as drainage canals, are built with a graded slope. Fish ponds should be built with an adequate bottom slope so that you will be able to drain them completely. Knowing how to set graded lines of slope is therefore very important when you are building a fish-farm.

2. You can set graded lines of slope in several ways, using **three series of methods** with the devices described in Chapters 4, 5 and 6.

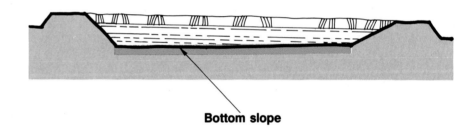

Bottom slope

Using slope-measuring devices for setting lines of slope

3. You can use any one of the **slope measuring devices described in Sections 41 to 46 to set graded lines of slope. The clisimeter, in particular (see Section 45), is commonly used for this purpose, but any other clinometer can be used instead.**

4. **It is best to use a target levelling staff** such as the one described in Section 65; its target should be tightly attached at your eye level. Remember that you can use the height of your assistant as a reference level instead.

5. From the starting point A of the line of slope, sight the target levelling staff; the **graduation of the clinometer** should correspond to the slope you have chosen. Signal to your assistant to move the levelling staff up or down the slope, until the sighting line of the clinometer lines up with the reference mark on the levelling staff. Mark the ground point B with a stake and repeat the procedure from that point.

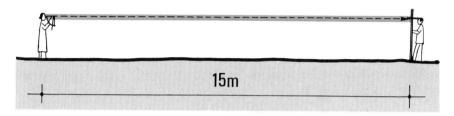

Use the target levelling staff

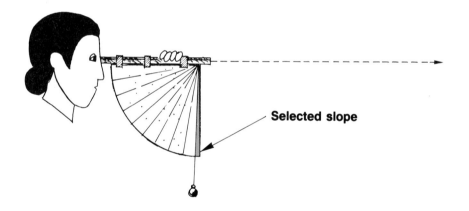

Selected slope

Sight with the clinometer at the desired graduation

Using sighting levels for setting lines of slope

6. You learned about various **sighting levels** in Sections 54 to 59 and in Section 65. These devices may be used to set lines of slope but, with the exception of the surveyor's level and the theodolite, their limited accuracy makes it difficult to lay slopes with gradients less than 1 percent. For smaller gradients, it is best to use non-sighting levels (see from step 12, below).

7. Before using the sighting level, calculate the **difference in height** (**H** metres) between two consecutive points according to their **horizontal distance** (**D** metres) in order to find the desired slope gradient (**S** percent) as:

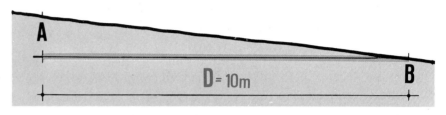

Measure the horizontal distance

$$H = (S \div 100) \times D$$

Example

- You decide to read levels at 10 m intervals, horizontal distance;
- The slope you need to set equals 1% or 1 m per 100 m;
- The necessary height difference H over a 10 m horizontal distance equals: (1 ÷ 100) × 10 m = 0.10 m.

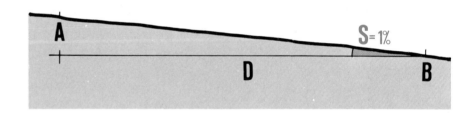

...and the slope

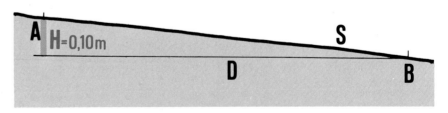

...to calculate the height difference

8. On the highest point A of the slope to be set, station your levelling device and measure the **height of its sighting line (H')** above the ground. Add this value to **H** (calculated in step 7) to obtain the **height to be read (R)** at the next point on the levelling staff as:

$$R = H + H'$$

9. Measure a horizontal distance of 10 m from the starting point, following the **contour line** as closely as possible. Place a levelling staff vertically at that point.

Note: for this part of the procedure, you can use:

- a **graduated levelling staff** on which you clearly mark the calculated height **R** (see step 8); or
- a target **levelling staff**, with the target tightly attached at the calculated height **R**.

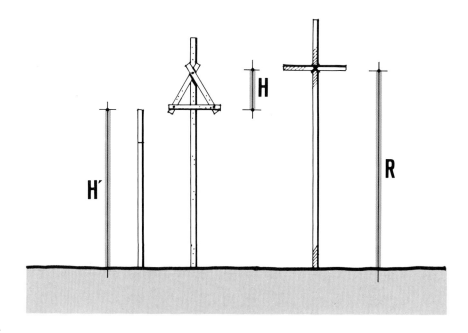

Set your target levelling staff to show the proper height difference

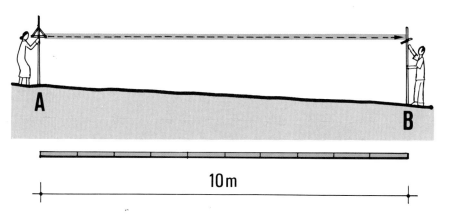

Place the levelling staff 10 m from the level

10. Sight with the levelling device at the levelling staff. Signal to your assistant to move the staff up or down the slope until the sighting line lines up with the mark **R** on the staff. At this point B, have your assistant drive a marking stake into the ground. Point B will be 10 cm lower than point A.

11. Station the levelling device on this marked point B while your assistant walks ahead another 10 m with the levelling staff. Repeat the procedure.

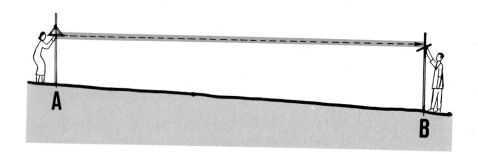

Move the levelling staff until the target comes into the line of sight

Proceed from point B

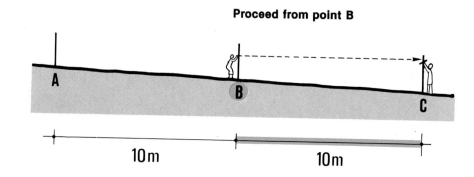

10 m   10 m

Using non-sighting levels for setting lines of slope

12.   **Non-sighting levels** are much more accurate than simple sighting levels for setting lines of slope with gradients smaller than 1 percent. Generally, non-sighting levels can be used to set lines of slope with **gradients as small as 0.3 percent**. The **flexible-tube water level** is even reliable for **slopes as small as 0.1 percent**.

13.   In Sections 62 to 64 and Section 66, you learned how to use various non-sighting levels for **contouring**, that is, setting lines of slope with a **zero gradient**.

14.   To set lines of slope with a different gradient (**S%**), you can use the same procedure described for contouring; the only difference is that you have to keep the forward end of the levelling device above the ground at the height H (calculated as shown above in step 7) for a fixed horizontal distance **D** metres, as in:

$$H = (S \div 100) \times D$$

Distance **D** varies according to the kind of levelling device you use.

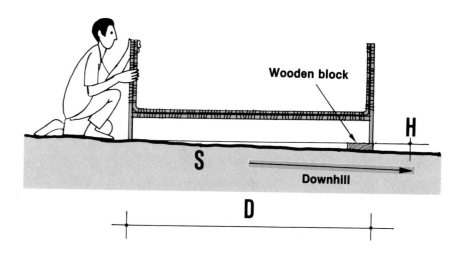

319

Note: it is best **to level going downhill**, as suggested above. If you must level going uphill, you should make the rear end of the levelling device higher by H metres.

15. The best way to do this is to prepare a **piece of wood with a thickness equal to H**. While levelling, place (or, better, nail) this piece of wood under the forward end of the level if you are levelling downhill.

Example

If S = 0.5 percent and you are levelling downhill:

- using an A-frame level for setting the line of slope:
  D = 4 m; **H** = 2 cm
- using an H-frame level:
  D = 2.5 m; **H** = 1.25 cm
- using a straight-edge level:
  D = 3 m; **H** = 1.5 cm
- using a flexible-tube water level:
  D = 10 m; **H** = 5 cm

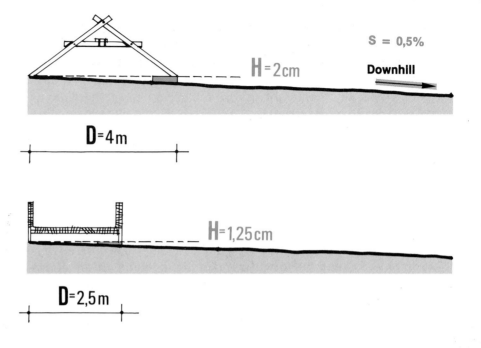

S = 0,5%

Downhill

H=2cm

D=4m

H=1,25cm

D=2,5m

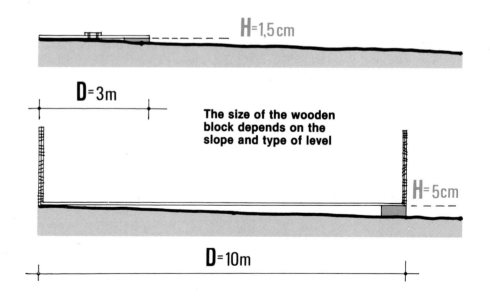

H=1,5cm

D=3m

The size of the wooden block depends on the slope and type of level

H=5cm

D=10m

16. When using the line level, you can add the height H to the cord height which will be maintained by the front person, instead of placing a piece of wood under the forward measuring staff.

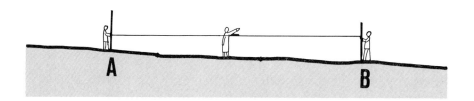

Example

If S = 0.5 percent and you are levelling downhill with a **line level:**
D − 20 m and **H** = 10 cm. **Keep the forward end** of the cord at a height 10 cm higher than the rear end of the cord.

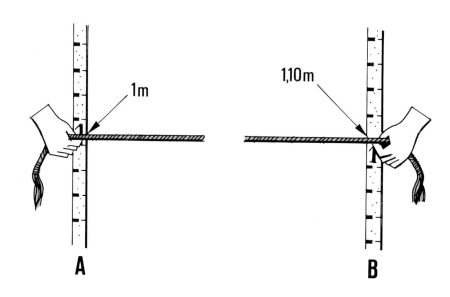

MEASUREMENT UNITS

Lengths/distances	**km**	= kilometre	= 1 000 m
Height differences	**m**	= metre	
Elevations	**cm**	= centimetre	= 0.01 m
	mm	= millimetre	= 0.001 m
Surface areas	**km²**	= square kilometre	= 100 ha
	ha	= hectare	= 10 000 m²
	m²	= square metre	
	cm²	= square centimetre	= 100 mm²
	mm²	= square millimetre	
Volumes	**m³**	= cubic metre	
Angles	**sec**	= ″ = second	
	min	= ′ = minute = 60 sec	
		° = degree = 60 min = 3 600 sec	
Slopes/gradients	**%**	= percent	
	‰	= per thousand	

COMMON ABBREVIATIONS

Az	=	azimuth (magnetic)	**FS**	=	foresight
BS	=	backsight	**HI**	=	height of the instrument
BM	=	bench-mark	**LS**	=	levelling station
Cos	=	cosine (angle)	**Sin**	=	sine (angle)
CI	=	contour interval	**Tan**	=	tangent (angle)
E(A)	=	elevation of point A	**TP**	=	turning point
			TBM	=	temporary bench-march

GLOSSARY OF TECHNICAL TERMS[1]

ALIDADE — Sighting ruler used together with a plane-table.

ALTITUDE — Vertical distance or height above the mean sea level which is in this case the reference **horizontal plane***; see also **"elevation"** and **"reference"**.

AZIMUTH — **Horizontal*** angle formed by **the magnetic north*** and a straight line or a direction; always measured clockwise from the magnetic north to the line/direction.

BACKSIGHT —
(a) Direction of a line measured when looking back to a previous survey point from a new point, whose direction has been defined as a foresight from the previous survey point. Commonly used in traversing.
(b) Measurement of the height above ground of a point of known elevation*, for example in direct levelling; in this case, also known as a plus sight*.

BENCH-MARK — Permanent, well-defined ground point of known or assumed elevation* used for example as the starting point of a topographical survey or as a reference point during constructions. A temporary bench-mark is used for a short period of time only, and is not permanently marked as a reference point.

CONTOUR — Imaginary line which joins all ground points of an equal elevation* above a given reference plane.

CONTOUR INTERVAL — Difference in elevation* between two adjacent contours*

CONTOUR LINE — A drawn line which joins points of equal elevation on a plan or a map; it represents a contour as it is found in the field.

---

[1] This glossary contains definitions of the technical terms marked with an asterisk (*) in the text.

CUMULATIVE DISTANCE
: Total distance from the starting point of a survey line.

CUT
: Area where it is necessary to lower the land to a required elevation*, by digging soil away.

ELEVATION
: **Vertical*** distance or height above a given **reference horizontal plane***; see also **"altitude"** and **"reference"**.

FILL
: Area where it is necessary to raise the land to a required elevation* by bringing soil in.

FORESIGHT
: (a) Direction of a line measured ahead (forward), from the line's initial point, for example in traversing.
: (b) Measurement of the height above ground of a point of unknown elevation*, for example in direct levelling; in this case, also known as a minus sight*.

GRAVITY
: Attractive force by which all bodies (including water) tend to move toward the centre of the earth, for example when moving or falling from a higher elevation to a lower elevation.

GROUND PROFILE
: A drawn representation of the ground surface which shows change in elevation* (along the vertical axis) with distance (along the horizontal axis).

HEIGHT OF THE INSTRUMENT
: Height above ground of the sightingline of a levelling* instrument.

HORIZONTAL
: Line or **plane***, **parallel*** to the plane of the horizon and at right angles to the **vertical plane***; flat, level.

LEVELLING
: Operation of measuring differences in elevation* at several ground points through a topographical survey.

GLOSSARY OF TECHNICAL TERMS, continued

LEVELLING STATION — Ground point where a levelling instrument is set up for a topographical survey.

LINE OF SIGHT — Imaginary line which begins at the eye of the surveyor and runs towards a fixed point; it is always a straight line, also called the **"sighting line"**.

MAGNETIC BEARING — Direction in which any point lies from a point of reference as measured from the **magnetic north*** with a compass.

MAGNETIC NORTH — Direction in which the magnetized end of a compass needle points ... towards the north magnetic pole of the earth. Note: the position of the north magnetic pole is affected by local variations, and corrections may be required for detailed use.

MINUS SIGHT — An elevation which is always subtracted, see foresight*, definition (b).

OBLIQUE LINE — For a given **horizontal** and **vertical plane***, an oblique line is:
— the **horizontal plane*** but not **perpendicular*** to the **vertical plane***, or
— within the **vertical plane***, but not **perpendicular*** to the **horizontal plane***, or
— within neither **plane***.

PARALLEL (LINE) — A line equally distant from another line at every point along its length.

PERPENDICULAR — A line/plane having a direction at right angles to a given line/plane.

PLANE — An imaginary flat surface; every straight line joining any two points in it lies totally in it.

PLUS SIGHT — An elevation which is always added, see backsight*, definition (b).

327

GLOSSARY OF TECHNICAL TERMS, continued

POINT OF REFERENCE Fixed point, usually identified in the field by a marker at the end of a line of sight.

POLYGON A geometrical figure or an area of land having more than three straight sides.

RECTANGLE A four-sided polygon* with four right angles*.

REDUCED LEVEL Vertical distance to a **common reference plane***, such as the mean sea level (see **"altitude"**) or an assumed horizontal plane (see **"elevation"**); it is calculated from survey data.

REFERENCE LEVEL/ PLANE **Elevation*** or **plane*** which is repeatedly used during a particular survey, and to which survey points or lines are referred.

RIGHT ANGLE A 90-degree angle.

RIGHT-ANGLED A triangle* with one 90-degree angle.

SIGHTING LINE Synonym for **"line of sight"**.

TANGENT Mathematical function for angles.

TRAPEZIUM A four-sided polygon* with two parallel* sides.

TRAVERSE A set of straight lines con necting established points around or along the route of a plan survey.

TRIANGLE A three-sided geometrical figure or land area.

TRIANGLE SCALE Relationship existing between the distance shown on a drawing and the actual distance across the ground.

TURNING POINT Temporary intermediate or reference point being surveyed between two established points; it is no longer needed after the necessary reading has been taken.

328

VERTICAL

Line or **plane*** which is perpendicular to a horizontal line or **plane***; in practice defined by the position of a freely suspended weighted line.

FURTHER READING

CLENDINNING, J. and J.G. OLLIVER, 1966. *The principles of surveying.* London, Blackie and Son Ltd., 463 p. 3rd. edition.

STERN, P. *et al.* (eds), 1983. *Field engineering. An introduction to development work and construction in rural areas.* London, Intermediate Technology Publications Ltd., 251 p.

NOTES

NOTES

NOTES

NOTES

WHERE TO PURCHASE FAO PUBLICATIONS LOCALLY
POINTS DE VENTE DES PUBLICATIONS DE LA FAO
PUNTOS DE VENTA DE PUBLICACIONES DE LA FAO

● **ALGÉRIE**
ENAMEP – Entreprise nationale des messageries de presse
47, rue Didouche Mourad, Alger.

● **ANGOLA**
Empresa Nacional do Disco e de Publicaçoes, ENDIPU-U.E.E.
Rua Cirilo de Conceiçao Silva, No. 7, C.P. No. 1314-C Luanda.

● **ARGENTINA**
Libreria Agropecuaria S.A.
Pasteur 743, 1028 Buenos Aires.

● **AUSTRALIA**
Hunter Publications
58A Gipps Street, Collingwood, Vic. 3066.

● **AUSTRIA**
Gerold & Co.
Graben 31, 1011 Vienna.

● **BAHRAIN**
United Schools International
PO Box 726, Manama.

● **BANGLADESH**
Association of Development Agencies in Bangladesh
1/3 Block F, Lalmatia, Dhaka 1209.

● **BELGIQUE**
M. J. De Lannoy
202, avenue du Roi, 1060 Bruxelles. CCP 000-0808993-13.

● **BOLIVIA**
Los Amigos del Libro
Perú 3712, Casilla 450, Cochabamba; Mercado 1315, La Paz.

● **BOTSWANA**
Botsalo Books (Pty) Ltd
PO Box 1532, Gaborone.

● **BRAZIL**
Fundação Getulio Vargas
Praia de Botafogo 190, C.P. 9052, Rio de Janeiro.
Livraria Canuto Ltda
Rua Consolação, 348 – 2º andar, Caixa Postal 19198, São Paulo.

● **BRUNEI-DARUSSALAM**
SST Trading Sdn. Bhd.
Bangunan Tekno No. 385, Jln 5/59, PO Box 227, Petaling Jaya, Selangor.

● **CANADA**
Renouf Publishing Co. Ltd
1294 Algoma Road, Ottawa, Ont. K1B 3W8.
Editions Renouf Ltée
route Transcanadienne, Suite 305, St-Laurent (Montréal), Qué.
Toll free calls: Ontario, Quebec and Maritime – 1-800-267-1805; Western Provinces and Newfoundland – 1-800-267-1826.
Head Office/Siège social: 1294 Algoma Road, Ottawa, Ont.

● **CHILE**
Libreria – Oficina Regional FAO
Avda. Santa Maria 6700, Casilla 10095, Santiago.
Teléfono: 228-80-56.

● **CHINA**
China National Publications Import Corporation
PO Box 88, Beijing.

● **CONGO**
Office national des librairies populaires
B.P. 577, Brazzaville.

● **COSTA RICA**
Libreria, Imprenta y Litografia Lehmann S.A.
Apartado 10011, San José.

● **CUBA**
Ediciones Cubanas, Empresa de Comercio Exterior de Publicaciones
Obispo 461, Apartado 605, La Habana.

● **CYPRUS**
MAM
PO Box 1722, Nicosia.

● **CZECHOSLOVAKIA**
ARTIA
Ve Smeckach 30, PO Box 790, 111 27 Prague 1.

● **DENMARK**
Munksgaard Export and Subscription Service
35 Nørre Søgade, DK 1370 Copenhagen K.

● **ECUADOR**
Libri Mundi, Libreria Internacional
Juan León Mera 851, Apartado Postal 3029, Quito.
Su Libreria Cia. Ltda.
Garcia Moreno 1172 y Mejia, Apartado Postal 2556, Quito.

● **EL SALVADOR**
Libreria Cultural Salvadoreña, S.A. de C.V.
7ª Avenida Norte 121, Apartado Postal 2296, San Salvador.

● **ESPAÑA**
Mundi-Prensa Libros S.A
Castelló 37, 28001 Madrid.
Libreria Agricola
Fernando VI 2, 28004 Madrid.

● **FINLAND**
Akateeminen Kirjakauppa
PO Box 128, 00101 Helsinki 10.

● **FRANCE**
Editions A. Pedone
13, rue Soufflot, 75005 Paris.

● **GERMANY, FED. REP.**
Alexander Horn Internationale Buchhandlung
Kirchgasse 39, Postfach 3340, 6200 Wiesbaden.
UNO Verlag
Poppelsdorfer Allee 55, D-5300 Bonn 1.
Triops Verlag
Raiffeisenstr. 24, 6060 Langen.

● **GHANA**
Ghana Publishing Corporation
PO Box 4348, Accra.

● **GREECE**
G.C. Eleftheroudakis S.A.
4 Nikis Street, Athens (T-126).
John Mihalopoulos & Son S.A.
75 Hermou Street, PO Box 73, Thessaloniki.

● **GUATEMALA**
Distribuciones Culturales y Técnicas "Artemis"
5ª Avenida 12-11, Zona 1, Apartado Postal 2923, Guatemala.

● **GUINEA-BISSAU**
Conselho Nacional da Cultura
Avenida da Unidade Africana, C.P. 294, Bissau.

● **GUYANA**
Guyana National Trading Corporation Ltd
45-47 Water Street, PO Box 308, Georgetown.

● **HAÏTI**
Librairie "A la Caravelle"
26, rue Bonne Foi, B.P. 111, Port-au-Prince.

● **HONDURAS**
Escuela Agricola Panamericana, Libreria RTAC
Zamorano, Apartado 93, Tegucigalpa.
Oficina de la Escuela Agricola Panamericana en Tegucigalpa
Blvd. Morazán, Apts. Glapson, Apartado 93, Tegucigalpa.

● **HONG KONG**
Swindon Book Co.
13-15 Lock Road, Kowloon.

● **HUNGARY**
Kultura
PO Box 149, 1389 Budapest 62.

● **ICELAND**
Snaebjörn Jónsson and Co. h.f.
Hafnarstraeti 9, PO Box 1131, 101 Reykjavik.

● **INDIA**
Oxford Book and Stationery Co.
Scindia House, New Delhi 100 001; 17 Park Street, Calcutta 700 016
Oxford Subscription Agency, Institute for Development Education
1 Anasuya Ave, Kilpauk, Madras 600010.

● **INDONESIA**
P.T. Inti Buku Agung
13 Kwitang, Jakarta.

● **IRAQ**
National House for Publishing, Distributing and Advertising
Jamhuria Street, Baghdad.

● **IRELAND**
Agency Section, Publications Branch
Stationery Office, Bishop Street, Dublin 8.

● **ITALY**
FAO (see last column)
Libreria Scientifica Dott. Lucio de Biasio "Aeiou"
Via Meravigli 16, 20123 Milan.
Libreria Commissionaria Sansoni S.p.A. "Licosa"
Via Lamarmora 45, C.P. 552, 50121 Florence.
Libreria Internazionale Rizzoli
Galleria Colonna, Largo Chigi, 00187 Rome.

● **JAPAN**
Maruzen Company Ltd
PO Box 5050, Tokyo International 100-31.

● **KENYA**
Text Book Centre Ltd
Kijabe Street, PO Box 47540, Nairobi.

● **KOREA, REP. OF**
Eulyoo Publishing Co. Ltd
46-1 Susong-Dong, Jongro-Gu, PO Box 362, Kwangwha-Mun, Seoul 110.

● **KUWAIT**
The Kuwait Bookshops Co. Ltd
PO Box 2942, Safat.

● **LUXEMBOURG**
M. J. De Lannoy
202, avenue du Roi, 1060 Bruxelles (Belgique).

● **MALAYSIA**
SST Trading Sdn. Bhd.
Bangunan Tekno No. 385, Jln 5/59, PO Box 227, Petaling Jaya, Selangor.

● **MAROC**
Librairie "Aux Belles Images"
281, avenue Mohammed V, Rabat.

● **MAURITIUS**
Nalanda Company Limited
30 Bourbon Street, Port-Louis.

● **NETHERLANDS**
Keesing b.v.
Hogeliweg 13, 1101 CB Amsterdam. Postbus 1118, 1000 BC Amsterdam.

● **NEW ZEALAND**
Government Printing Office Bookshops
25 Rutland Street.
Mail orders: 85 Beach Road, Private Bag, CPO, Auckland; Ward Street, Hamilton; Mulgrave Street (Head Office), Cubacade World Trade Centre, Wellington; 159 Hereford Street, Christchurch; Princes Street, Dunedin.

● **NICARAGUA**
Libreria Universitaria, Universidad Centroamericana
Apartado 69, Managua.

● **NIGERIA**
University Bookshop (Nigeria) Limited
University of Ibadan, Ibadan.

● **NORWAY**
Johan Grundt Tanum Bokhandel
Karl Johansgate 41-43, PO Box 1177, Sentrum, Oslo 1.

● **PAKISTAN**
Mirza Book Agency
65 Shahrah-e-Quaid-e-Azam, PO Box 729, Lahore 3.
Sasi Book Store
Zaibunnisa Street, Karachi.

● **PARAGUAY**
Agencia de Librerias Nizza S.A.
Casilla 2596, Eligio Ayala 1073, Asunción.

● **PERU**
Libreria Distribuidora "Santa Rosa"
Jirón Apurimac 375, Casilla 4937, Lima 1.

● **POLAND**
Ars Polona
Krakowskie Przedmiescie 7, 00-068 Warsaw.

● **PORTUGAL**
Livraria Portugal, Dias y Andrade Ltda.
Rua do Carmo 70-74, Apartado 2681, 1117 Lisbonne Codex.

● **REPUBLICA DOMINICANA**
Editora Taller, C. por A.
Isabel la Católica 309, Apartado de Correos 2190, ZI Santo Domingo.
Fundación Dominicana de Desarrollo
Casa de las Gàrgolas, Mercedes 4, Apartado 857, ZI Santo Domingo.

● **ROMANIA**
Ilexim
Calea Grivitei No 64066, Bucharest.

● **SAUDI ARABIA**
The Modern Commercial University Bookshop
PO Box 394, Riyadh.

● **SINGAPORE**
MPH Distributors (S) Pte. Ltd
71/77 Stamford Road, Singapore 6.
Select Books Pte. Ltd
215 Tanglin Shopping Centre, 19 Tanglin Rd., Singapore 1024.

● **SOMALIA**
"Samater's"
PO Box 936, Mogadishu.

● **SRI LANKA**
M.D. Gunasena & Co. Ltd
217 Olcott Mawatha, PO Box 246, Colombo 11.

● **SUDAN**
University Bookshop, University of Khartoum
PO Box 321, Khartoum.

● **SUISSE**
Librairie Payot S.A.
107 Freiestrasse, 4000 Basel 10.
6, rue Grenus, 1200 Genève.
Case Postale 3212, 1002 Lausanne.
Buchhandlung und Antiquariat Heinimann & Co.
Kirchgasse 17, 8001 Zurich.

● **SURINAME**
VACO n.v. in Suriname
Domineestraat 26, PO Box 1841, Paramaribo.

● **SWEDEN**
Books and documents:
C.E. Fritzes Kungl. Hovbokhandel,
Regeringsgatan 12, PO Box 16356, 103 27 Stockholm.
Subscriptions:
Vennergren-Williams AB
PO Box 30004, 104 25 Stockholm.

● **TANZANIA**
Dar-es-Salaam Bookshop
PO Box 9030, Dar-es-Salaam.
Bookshop, University of Dar-es-Salaam
PO Box 893, Morogoro.

● **THAILAND**
Suksapan Panit
Mansion 9, Rajadamnern Avenue, Bangkok.

● **TOGO**
Librairie du Bon Pasteur
B.P. 1164, Lomé.

● **TUNISIE**
Société tunisienne de diffusion
5, avenue de Carthage, Tunis.

● **TURKEY**
Kultur Yayinlari Is-Turk Ltd Sti.
Ataturk Bulvari No. 191, Kat. 21, Ankara
Bookshops in Istanbul and Izmir.

● **UNITED KINGDOM**
Her Majesty's Stationery Office
49 High Holborn, London WC1V 6HB (callers only).
HMSO Publications Centre, Agency Section
51 Nine Elms Lane, London SW8 5DR (trade and London area mail orders);
13a Castle Street, Edinburgh EH2 3AR;
80 Chichester Street, Belfast BT1 4JY;
Brazennose Street, Manchester M60 8AS;
258 Broad Street, Birmingham B1 2HE;
Southey House, Wine Street, Bristol BS1 2BQ.

● **UNITED STATES OF AMERICA**
UNIPUB
4611/F, Assembly Drive, Lanham, MD 20706.

● **URUGUAY**
Libreria Agropecuaria S.R.L.
Alzaibar 1328, Casilla Correo 1755, Montevideo.

● **YUGOSLAVIA**
Jugoslovenska Knjiga, Trg.
Republike 5/8, PO Box 36, 11001 Belgrade.
Cankarjeva Zalozba
PO Box 201-IV, 61001 Ljubljana.
Prosveta
Terazije 16, Belgrade.

● **ZAMBIA**
Kingstons (Zambia) Ltd
Kingstons Building, President Avenue, PO Box 139, Ndola.

Other Countries
Autres Pays
Otros Paises

Distribution and Sales Section, FAO
Via delle Terme di Caracalla, 00100 Rome, Italy.